Statistics for
Six Sigma Black Belts

Also available from ASQ Quality Press:

The ASQ Pocket Guide to Statistics for Six Sigma Black Belts
Matthew A. Barsalou

The Certified Six Sigma Black Belt Handbook, Third Edition
T.M. Kubiak and Donald W. Benbow

Practical Engineering, Process, and Reliability Statistics
Mark Allen Durivage

Reliability Data Analysis with Excel and Minitab
Kenneth S. Stephens

Zero Acceptance Number Sampling Plans, Fifth Edition
Nicholas L. Squeglia

The Certified Quality Engineer Handbook, Third Edition
Connie M. Borror, editor

The Quality Toolbox, Second Edition
Nancy R. Tague

Root Cause Analysis: Simplified Tools and Techniques, Second Edition
Bjørn Andersen and Tom Fagerhaug

The Certified Six Sigma Green Belt Handbook, Second Edition
Roderick A. Munro, Govindarajan Ramu, and Daniel J. Zrymiak

The Certified Manager of Quality/Organizational Excellence Handbook, Fourth Edition
Russell T. Westcott, editor

The ASQ Auditing Handbook, Fourth Edition
J.P. Russell, editor

The ASQ Quality Improvement Pocket Guide: Basic History, Concepts, Tools, and Relationships
Grace L. Duffy, editor

To request a complimentary catalog of ASQ Quality Press publications, call 800-248-1946, or visit our website at http://www.asq.org/quality-press.

Statistics for Six Sigma Black Belts

Matthew A. Barsalou

ASQ Quality Press
Milwaukee, Wisconsin

American Society for Quality, Quality Press, Milwaukee 53203
© 2015 by ASQ
All rights reserved.
Printed in the United States of America
20 19 18 17 16 15   5 4 3 2 1

Library of Congress Cataloging-in-Publication Data
Barsalou, Matthew A., 1975–
   Statistics for six sigma black belts / Matthew A. Barsalou.
     pages cm
   Includes bibliographical references and index.
   ISBN 978-0-87389-892-8 (alk. paper)
   1. Six sigma (Quality control standard) 2. Quality control—Statistical methods. I. Title.
effectiveness. I. Title.
  TS156.B4316 2015
  658.5′620218—dc23

                                                     2014038327

Publisher: Lynelle Korte
Acquisitions Editor: Matt Meinholz
Managing Editor: Paul Daniel O'Mara
Production Administrator: Randall Benson

ASQ Mission: The American Society for Quality advances individual, organizational, and community excellence worldwide through learning, quality improvement, and knowledge exchange.

Attention Bookstores, Wholesalers, Schools, and Corporations: ASQ Quality Press books, video, audio, and software are available at quantity discounts with bulk purchases for business, educational, or instructional use. For information, please contact ASQ Quality Press at 800-248-1946, or write to ASQ Quality Press, P.O. Box 3005, Milwaukee, WI 53201-3005.

To place orders or to request a free copy of the ASQ Quality Press Publications Catalog, visit our website at http://www.asq.org/quality-press.

 Printed on acid-free paper

Quality Press
600 N. Plankinton Ave.
Milwaukee, WI 53203-2914
E-mail: authors@asq.org
The Global Voice of Quality™

Dedicated to my father, Gilbert L. Barsalou, my brother, Mark A. Barsalou, and two dear friends whom I lost while writing this, Michael W. "Pugsly" Adams and Sarah Jung.

Table of Contents

List of Figures and Tables

Preface

Just over half a decade before the arrival of Six Sigma, William J. Diamond (1981) wrote, "The best experiment designs result from the combined efforts of a skilled experimenter, who has had basic training in experiment design methods, and of a skilled statistician, who has had training in engineering or science. The statistician alone cannot design good experiments in every possible discipline; neither can the scientist or engineer who is untrained in statistical experiment design be a good experiment designer." Today, Six Sigma Black Belts are expected to have the skills of a good experimenter, possessing both a deep understanding of statistics and a knowledge of the industry in which they work.

This does not mean a Six Sigma Black Belt must know everything; the Six Sigma project team should include experts with the required detailed technical knowledge of the process being improved. A Six Sigma Black Belt can also consult with a Master Black Belt or a statistician for additional support with statistical methods.

Statistics for Six Sigma Black Belts is written for the Six Sigma Black Belt who needs an understanding of many statistical methods but does not use all of these methods every day. A Six Sigma Black Belt who has not had to use a specific statistical test in several years should be able to quickly review the test and perform it using the examples presented here. This book is intended to be used as a quick reference, providing basic details as well as step-by-step instructions for using Minitab® statistical software.

Six Sigma Black Belts typically use a statistical program to perform calculations, but an understanding of the underlying statistics is still needed. Anybody can type data into a program; a Black Belt must be capable of understanding which hypothesis test is appropriate for a given use as well as the assumptions that must be met to correctly perform the hypothesis test.

The methods presented here are laid out according to the Six Sigma DMAIC (Define, Measure, Analyze, Improve, Control) phases in which they are typically used. However, these methods can also be applied outside of a Six Sigma project, such as when one simply needs to determine whether there is a difference in the means of two processes producing the same parts. In such a case, the flowchart in Appendix A could be used to quickly identify the correct test based on the intended use and available data.

A Six Sigma Black Belt using *Statistics for Six Sigma Black Belts* will be able to quickly zero in on the appropriate method and follow the examples to reach the correct statistical conclusion.

ACKNOWLEDGMENTS

This book would not have been possible without the support of many people. I would like to thank Lynetta Campbell (lkcampbell@aol.com) for keeping my statistics straight. I would also like to thank Jim Frost from Minitab for his assistance in providing clear conclusions for hypothesis tests, and Eston Martz and Michelle Paret from Minitab for providing me with a DOE data set. I am also grateful to Dean Christolear of CpkInfo.com for letting me use his Z-score table, and Rick Haynes of Smarter Solutions (www.smartersolutions.com) for providing me with the templates to create the other statistical tables.

Portions of information contained in this book are printed with permission of Minitab, Inc. All such material remains the exclusive property and copyright of Minitab, Inc. All rights reserved.

Introduction

Six Sigma projects have five phases: Define, Measure, Analyze, Improve, and Control. The statistical methods presented here are laid out according to the phase in which they are typically used. This book is intended to present the statistics of Six Sigma; however, it would be negligent to fail to mention the phases in which the methods are applied.

Chapter 1 briefly covers the Define phase. Chapter 2 covers the Measure phase, where baseline performance is determined. Here, basic statistical concepts as well as types of data and measurement scales are introduced. Samples and populations are described and basic probability is explained. The binomial distribution and Bayes' theorem are described. The chapter then covers descriptive statistics including measures of central tendency, variability, and shape before going into process capability and performance and measurement system analysis.

Chapter 3 deals with methods used during the Analyze phase of a Six Sigma project. Hypothesis testing is detailed, including error types and the five steps for hypothesis testing. Hypothesis tests presented include Z-tests, t-tests, and tests for both population and sample proportions. Also included are confidence intervals for means, the chi-square test of population variance, and the F-test of the variance of two samples, as well as simple linear regression and analysis of variance (ANOVA) for testing the differences between means.

Design of experiments (DOE) for use during the Improve phase of a Six Sigma project is detailed in Chapter 4. Chapter 5, which covers the Control phase, presents statistical process control (SPC), including the I-mR chart, the $\bar{x}$ and S chart, and the $\bar{x}$ and R chart for variable data, and the c chart, u chart, p chart, and np chart for attribute data.

Appendix A includes a flowchart that provides the correct statistical test for a given use and type of data. Also included are flowcharts depicting the five steps for hypothesis testing. Appendix B contains the statistical formulas in tables to serve as a quick reference. A Black Belt who only needs to look up a formula can find it in the tables; if the Black Belt is unsure how to apply the formula, he or she can follow the instructions in the appropriate chapter. The statistical tables are located in Appendix C. These include tables for the Z score, t distribution, F distribution, and chi-square distribution. Appendix D provides the SPC constants. Appendix E is a quick reference for those unfamiliar with Minitab. A detailed glossary is included as Appendix F.

Chapter 1
Define

The Define phase is the first phase of a Six Sigma project. Typical Define phase activities include creating a project charter that lays out the project's goals and timeline and a clear project statement (George et al. 2005). The project statement should clearly communicate the problem and its impact on the business, and it should include goals and strategic objectives. It should also help the project team to focus on the core issue (Breyfogle 2008). The scope of the project should also be determined; this involves defining what is part of the project and what is clearly outside the bounds of the project.

Project management tools such as Gantt and PERT charts are often created during the Define phase. These are used to identify and track project milestones. An activity tracking list may also be created at this time to track the status of delegated project tasks.

Chapter 2
Measure

The current level of performance is assessed during the Measure phase of a Six Sigma project. Process variables are identified using tools such as SIPOC (supplier-input-process-output-customer), and flowcharts are used to better understand the process that is being improved. Often, work and process instructions will be checked to determine how the process is described. A failure modes and effects analysis (FMEA) may be created for a design concept or a process.

The $Y = f(x)$ is also established during the Measure phase (Shankar 2009); this is the critical factor or factors that must be controlled in order for a process to function properly. For a Six Sigma project seeking to reduce delivery times, the factors influencing delivery time must be identified. For a machining process, the $Y = f(x)$ may be settings on the machine that result in the desired surface finish.

The baseline performance is often established in terms of parts per million (PPM) defective or defects per million opportunities (DPMO). This baseline measurement is helpful in determining whether improvement efforts were successful. To determine PPM, simply divide the number of defective parts by the total number of parts and multiply by 1 million:

$$\text{PPM} = \frac{\text{number of defective parts}}{\text{total number of parts}} \times 1{,}000{,}000$$

DPMO looks at the number of defects instead of defective parts; one part may have multiple defects. To determine DPMO, multiply the number of defects by 1 million and divide by the total number of parts times the number of opportunities for one part to have a defect:

$$\text{DPMO} = \frac{1{,}000{,}000 \times \text{number of defects}}{\text{total number of parts} \times \text{number of opportunities per part}}$$

The same calculation can be applied to a business process such as invoicing:

$$\text{DPMO} = \frac{1{,}000{,}000 \times \text{number of mistakes in invoices}}{\text{total number of unique data points} \times \text{number of invoices}}$$

Care must be used when determining what constitutes a defect, or the resulting DPMO figure could be unrealistic. It would be realistic to view an assembly process with five distinct steps as having five opportunities for a defect to occur; it would be unrealistic to count every movement the assembly operator makes as an opportunity for a defect to occur. Using an unrealistic number of potential defect opportunities will make a process appear better than it really is.

Basic statistical methods are used during the Measure phase to gain a better understanding of the performance of the process. Measuring devices that will be used during the Six Sigma project are assessed to ensure they consistently deliver correct results, and capability studies are performed to determine the current level of process performance.

TYPES OF DATA AND MEASUREMENT SCALES

Data can be either qualitative or quantitative (see Table 2.1). Qualitative data are measures of types and are often represented by names or descriptions. Qualitative data may be represented by numbers, such as when a researcher codes females as a 1 and males as a 2; however, the data are still qualitative in spite of the use of a number. Quantitative data are measures of values or counts and are expressed as numbers. Qualitative data can be thought of as dealing with descriptions, such as red, old, or soft. Quantitative data are measurable; length, weight, and temperature are examples of quantitative data.

Quantitative data can be continuous or discrete, while qualitative data are always discrete. Continuous data, also known as variable data, have unlimited possibilities—time or length, for example—and have no clear boundaries between nearby values. Discrete data, also known as attribute data, are countable—the number of defective parts, for example—and they have clear boundaries and include counts and rank orders. Discrete data can consist of yes/no data or individual whole values, such as the number of defective parts (Pries 2009).

There are also different types of measurement scales, such as nominal, ordinal, interval, and ratio (see Table 2.2). Nominal data consist of names (e.g., "machine one") as well as data such as part numbers. Nominal data cannot be continuous because they are strictly qualitative; the numbers serve as labels with no numerical meaning attached. Ordinal, interval, and ratio are all ways to represent quantitative variables. Ordinal data show the relative place of the data and use unequal value ranking, such as first place and second place. Interval data have equal distances between data points, such as temperature in degrees Celsius or calendar

Table 2.1 Examples of qualitative and quantitative data.

Data type	Examples
Qualitative	Bolts (metric, standard), sheets (heavy, light), employee status (full time, part time), colors (green, blue), defect types (length deviation, rust)
Quantitative Discrete/Attribute Continuous/Variable	19 defects, 12 employees, 3 customer complaints Diameter of 0.25 centimeters, duration of 4 hours, speed of 60 kilometers per hour, turning speed of 750 revolutions per minute

Table 2.2 Examples of measurement scales.

Scale type	Examples
Nominal	Milling machine one, Part number 452-374c
Ordinal	First place, second place, largest, smallest, defective or not defective, somewhat agree or completely agree
Interval	Degrees Celsius, calendar month, IQ scores
Ratio	Weight in kilograms, length in centimeters

years, and may have an arbitrary zero point. Ratio data are used for proportional differences and can be compared to each other (Johnson 1988). Ratio data have a non-arbitrary zero value—for example, a 42 millimeter long shaft is twice as long as a 21 millimeter long shaft.

SAMPLES AND POPULATIONS

A population consists of all observations that might occur. Theoretically, a population could be infinite in size. A sample is the observation of a subset of the population (Box, Hunter, and Hunter 2005). Suppose a Six Sigma Black Belt wanted to study the length of rods cut on a machine. Theoretically, the machine could produce tens of thousands of parts and could do so over the course of weeks. The Six Sigma Black Belt selects nine parts from material that has been cut on the machine; these nine parts are the sample out of the population. A large population is shown in Figure 2.1; beneath the population is a smaller sample size of 16 that has been drawn from the population.

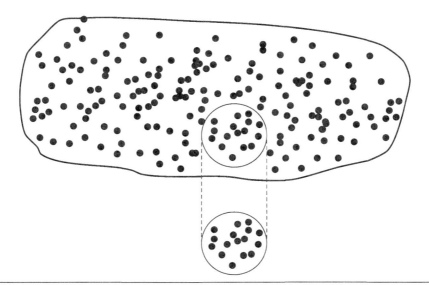

Figure 2.1 Sample drawn from a population.

PROBABILITY

Probability is based on sample space, which is all combinations of events (Goldberg 1986). We can calculate the probability of an event occurring if we know the probability of multiple individual events that must occur in order for the event of interest to happen.

The symbol "∪" is the symbol of a union and means "or." For example, $A{\cup}B$ represents the union of A and B, meaning all events that are included in A, B, or both. The union of A and B is shown in Figure 2.2. The symbol "∩" signifies an intersection and can be understood as implying "and." For example, $A{\cap}B$ represents the intersection of A and B, meaning only those events that occur in both A and B. The symbol "~" is one of several symbols that signify the complement of an event; therefore, $\sim A$ means that A has not occurred. A conditional probability is the probability that one event will occur given that another has occurred; it is written as $P(A\,|\,B)$, which reads as the probability that A will occur given that B has occurred.

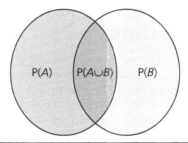

Figure 2.2 The union of A and B.

Addition Rule for Union of Two Events That Are Not Mutually Exclusive

Mutually exclusive occurrences are occurrences that can be one thing or the other, but not both. An example of this is a switch that can be in either the "on" or "off" position—not both at the same time. Vehicles with gasoline engines and vehicles with electric motors are mutually exclusive; a hybrid vehicle with both a gasoline engine and an electric motor is not mutually exclusive. The addition rule requires adding the two probabilities together and subtracting the probability of the intersection of the two events:

$$P(A{\cup}B) = P(A) + P(B) - P(A{\cap}B)$$

Example

Suppose we have a chess set with 16 black pieces and 16 white pieces. Half of the pieces are pawns. We want to determine the chance of randomly selecting a white piece or a pawn.

Procedure

Step 1: Assign P(A) to the white piece and P(B) to the pawn.

Step 2: To simplify the example, rewrite the equation using the chess pieces in place of A and B and replacing ∪ and ∩ with "or" and "and": P(white piece or pawn) = P(white piece) + P(pawn) – P(white piece and pawn).

Step 3: Calculate P(white piece), which is 16 / 32 = 0.5.

Step 4: Calculate P(pawn), which is 16 / 32 = 0.5.

Step 5: Determine P(white piece and pawn), which is 8 / 32 = 0.25.

Step 6: Set up the equation P(white piece or pawn) = 0.5 + 0.5 – 0.25 = 0.75.

Conclusion

There is a 75% chance of drawing a chess piece that is white, a pawn, or a white pawn.

Addition Rule for Union of Two Mutually Exclusive Events

As discussed in the previous section, mutually exclusive events must be one thing or the other; they cannot be both. For example, color is a mutually exclusive condition if something can be completely black or completely red but not completely black and completely red. The probabilities of each event are added to determine the probability that either one event or the other will occur. The formula is:

$$P(A \cup B) = P(A) + P(B)$$

Example

Suppose a bowl contains 10 red balls, 15 black balls, and 20 orange balls. We want to determine the probability of drawing a red ball or a black ball.

Procedure

Step 1: Assign P(A) to red and P(B) to black.

Step 2: To simplify the example, rewrite the equation using the colors in place of A and B and replacing ∪ with "or": P(red or black) = P(red) + P(black).

Step 3: Determine the probability of drawing a red ball, P(red). This is the number of red balls divided by the total number of balls, so 10 / 45 = 0.222.

Step 4: Determine the probability of drawing a black ball, P(black). This is the number of black balls divided by the total number of balls, so 15 / 45 = 0.333.

Step 5: Enter the probabilities into the formula, P(red or black) = P(red) + P(black). This is P(red or black) = 0.222 + 0.333 = 0.556.

Conclusion

There is a 55.6% chance of drawing a red ball or a black ball.

Multiplication Rule for the Intersection of Two Dependent Events

Events are dependent if they affect each other. For example, drawing a white pawn from a chess set affects the chance of drawing any other piece on the next draw if the pawn is not returned. If two events are dependent, the probability of them both occurring is written as $P(A \cap B) = P(A) \times P(B \mid A)$.

Example

A chess set contains a total of 32 pieces, which are placed in a bowl. There are 8 white pawns and 8 black pawns. What is the chance of drawing a white pawn from the bowl first and then a black pawn? The sampling is done without replacement, meaning the first piece is not returned to the bowl; otherwise the two events would not be dependent.

Note: This example ignores the selection bias. Each type of chess piece has a different shape, so a person reaching into a bowl of chess pieces might have a greater chance of drawing a smaller or larger piece depending on how the different sizes lie within the bowl. The person may also consciously or subconsciously select one type of piece over another.

Procedure

Step 1: Assign $P(A)$ to the drawing of a white pawn and $P(B)$ to the drawing of a black pawn.

Step 2: Rewrite the formula using the names of the chess pieces: P(white pawn and black pawn) = P(white pawn) × P(black pawn after white pawn is drawn). *Note:* This step is not necessary but could provide clarity for a complicated problem.

Step 3: Plug the probabilities into the formula and calculate: P(white pawn and black pawn) = 8 / 32 × 8 / 31 = 0.25 × 0.258 = 0.065.

Conclusion

The probability of drawing a white and then a black pawn is 0.065. The removal of the first piece automatically increases the chance of drawing the second piece from 8 out of 32 to 8 out of 31.

Multiplication Rule for the Intersection of Two Independent Events

Independent events are unrelated to each other. For example, the chance of rolling a 1 on a die is unrelated to the chance of rolling a 6 on another die because the dice rolls are unrelated to each other. If the chance of event A is unrelated to the chance of event B, then the probability of A is multiplied by the probability of B to determine the chance of both events happening. This is expressed mathematically as:

$$P(A \cap B) = P(A) \times P(B)$$

Naturally, the probability of both independent events occurring is much lower than the chance of either individual event occurring.

Example

Suppose there is a 9% chance of machine A breaking down during any given week and a 14% chance of machine B breaking down during any given week. What is the chance that both machines will break down during the same week?

Procedure

Step 1: Assign P(*A*) to machine A and P(*B*) to machine B.

Step 2: Rewrite the formula using the names of the machines: P(machine A and machine B breakdown) = P(machine A breakdown) × P(machine B breakdown). *Note:* This step is not necessary but could provide clarity for a complicated problem.

Step 3: Plug the probabilities into the formula and calculate: P(machine A and machine B breakdown) = 0.09 × 0.14 = 0.0126.

Conclusion

There is a 1.26% chance of both machines breaking down during the same week.

BINOMIAL DISTRIBUTION

The binomial distribution can be used to estimate binomial probabilities for k out of n situations, where k is the number of successes and n is the number of trials. Figure 2.3 shows a binomial distribution for 10 trials with an event probability of 0.1 and 10 trials with an event probability of 0.4. The graph shows that when P = 0.1 (i.e., the probability of success on each trial is 0.1), the probability of getting 0 successes in 10 trials is pretty high—about 0.35. When the probability of success on each trial increases to 0.4, the probability of getting 0 successes out of 10 trials is much lower.

The probability of success on any given trial is P; an exclamation point after a number means the factorial must be computed by multiplying the number preceded by an exclamation point by every lower positive integer. Unfortunately, such numbers can be very large, so it is advisable to use an online binomial calculator or a statistical program for the calculation. The number of trials must be set before performing the calculation, and each trial must be independent of the other

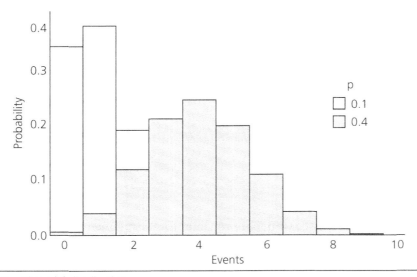

Figure 2.3 Binomial distribution.

trials. The binomial distribution may be one of the most important distributions in statistics (Clarke and Cooke 1992).

$$p(k) = \frac{n!}{k!(n-k)!} P^k (1-P)^{n-k}$$

Example

Suppose an engineer is going on a 10-day vacation and there is a 7% chance of rain each day. What is the chance that it will rain on exactly two days? *Note:* Rain on one day is assumed to be independent of rain on a second day for this simplified example.

Procedure

Step 1: Assign values to the letters in the formula and calculate the factors:

$n = 10$

$n! = 10 \times 9 \times 8 \times 7 \times 6 \times 5 \times 4 \times 3 \times 2 \times 1 = 3{,}628{,}800$

$k = 2$

$k! = 2 \times 1 = 2$

$n - k = 8$

$(n - k)! = 8 \times 7 \times 6 \times 5 \times 4 \times 3 \times 2 \times 1 = 40{,}320$

$P = 0.07$

Step 2: Plug the values into the formula and calculate.

$$P(\text{rain on 2 days}) = \frac{10!}{2!(10-2)!} 0.07^2 (1 - 0.07)^{10-2} =$$

$$\frac{3{,}628{,}800}{2(40{,}320)} 0.07^2 (1 - 0.07)^{10-2} = 0.123$$

Conclusion

There is a 12.3% chance that it will rain on exactly two days. Performing the same calculation for $k = 1$ results in a 36.4% chance of rain on exactly one day during the 10 days.

BAYES' THEOREM

Bayes' theorem uses conditional probabilities. It is frequently used in the medical field to determine the chance of somebody actually having a disease if they test positive for the disease (Boslaugh and Watters 2008). Instead of using distributions like many other statistical tests, Bayes' theorem uses known probabilities.

$$P(A \mid B) = \frac{P(A)P(B \mid A)}{P(B \mid A)P(A) + P(B \mid \sim A)P(\sim A)}$$

$P(A)$ is the chance of A occurring. $P(A \mid B)$ is the chance of A occurring given that B has occurred. $P(B \mid A)$ is the chance of B occurring given that A has occurred. $P(\sim A)$ is the chance of A not occurring. $P(B \mid \sim A)$ is the chance that B will occur given that A has not occurred.

Example

Suppose there is a 0.8% chance of an individual having a particular disease. A test has a 98% chance of diagnosing the disease when the disease is actually present, but it also has a 5% chance of diagnosing the disease when it is not present (known as a false positive). What is the chance of an individual having the disease when the test result is positive? What is the chance of an individual not having the disease when the test result is positive?

Procedure

Step 1: Find the probability for $P(A)$. An individual's chance of having the disease is given as 0.8%, which is converted from a percentage to a decimal as 0.008 to perform the calculations.

Step 2: Find the probability for $P(B|A)$, which is the chance of a test being positive when a person has the disease. This is given as 0.98.

Step 3: Find the probability for $P(B|\sim A)$, which is the chance of a test being positive when the disease is not present. It is given as 0.05.

Step 4: Determine the probability for $P(\sim A)$, which is the probability that a person does not have the disease. This is 1 minus the probability of a person having the disease. The chance of not having the disease is $1 - 0.008 = 0.992$.

Step 5: Determine the probability for $P(B)$, which is the chance of a test result being positive regardless of whether the disease is present. To calculate $P(B)$, multiply the probability of a test result being positive when the disease is present by the probability of a person having the disease and add the probability of a false positive times the probability that the disease is not present. The formula is $P(B|A) \times P(A) + P(B|\sim A) \times P(\sim A)$, so $0.98 \times 0.008 + 0.05 \times 0.992 = 0.0574$.

Step 6: Determine the probability for $P(\sim A|B)$, which is the chance of an individual getting a positive test result when the disease is not present (a false positive). This is determined by multiplying the chance of a test yielding a false positive by the chance of an individual not having the disease and dividing the resulting product by the total chance of a test result being positive. The formula is $(P(B|\sim A) \times P(\sim A)) / P(B)$, so $(0.05 \times 0.992) / 0.0574 = 0.864$; therefore, for any given person there is an 86.4% chance of getting a false positive.

Step 7: Determine the probability for $P(A|B)$, which is the probability that a person actually has the disease when the test result is positive. This requires Bayes' theorem; however, the individual elements have already been calculated. Steps 1 and 2 give us the two parts of the numerator, $P(A)$ and $P(B|A)$. The elements of the denominator can be found in steps 2, 1, 3, and 4; these are P($B|A$), P(A), P($B|\sim A$), and P($\sim A$). The denominator has already been calculated in step 5; therefore, use P(A) times P($B|A$) divided by P(B). This is 0.008 × 0.98 / 0.0574 = 0.1365; therefore, there is a 13.65% chance of a person having the disease when the test result is positive.

Conclusion

Using Bayes' theorem we have determined that due to the low occurrence rate and the 5% chance of a test yielding a false positive, the chance of a person actually having the disease when the test result is positive is only 13.65%. The chance of an individual getting a false positive test result is 86.4%. Therefore, a positive test result is not proof that a person has the disease and the person should be retested.

DESCRIPTIVE STATISTICS

Descriptive statistics are used to describe and summarize the basic features of data. Summarizing descriptive statistics is generally one of the first steps performed in any data analysis. This is in contrast to inferential statistics, which are used to extrapolate beyond the data set. An understanding of descriptive statistics is often needed to use inferential statistics; for example, performing a hypothesis test on the mean of two data samples may require calculating the standard deviation prior to performing the hypothesis test. Elements of descriptive statistics include measures of central tendency and measures of dispersion.

Measures of Central Tendency

Measures of central tendency describe the way in which data cluster around a numerical value. Measures of central tendency let us describe a distribution in terms of the center of the distribution (Johnson 1988). Measures of central tendency are the mean, the mode, and the median. The mean is the arithmetic average of a data set. This is calculated by adding together all the numbers in the data set and then dividing by n, where n is the number of items in the data set. The mode is simply the most frequently occurring number in a data set; there may be more than one. The median is the middle of the data set when the numbers are arranged in order from lowest to highest. Half of all data values fall above the median and half fall below the median. If the data set contains an even number of items, to determine the median it is necessary to add the two middle numbers and divide the resulting answer by 2.

The standard normal distribution has a bell shape and the mean, mode, and median are equal to each other, as shown in Figure 2.4. A distribution may also be skewed to the right, as shown in Figure 2.5, or to the left, as shown in Figure 2.6. A skew to the right is called a positive skew and a skew to the left is called a negative skew. Income is a good example of a distribution that would be skewed to the right; most people would be concentrated on the left side of the distribution, but the presence of millionaires and billionaires would pull the upper tail far out to the right.

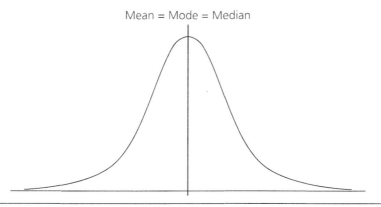

Figure 2.4   Normal distribution with mean = mode = median.

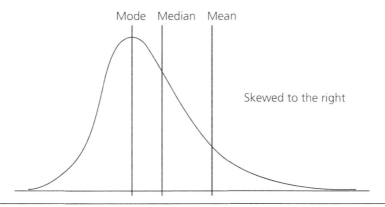

Figure 2.5   Right-skewed distribution.

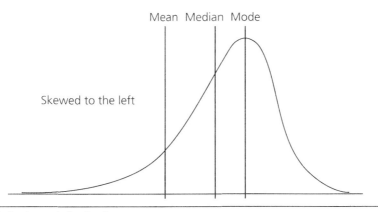

Figure 2.6   Left-skewed distribution.

Mean

The formula for the sample mean is $\bar{x} = \frac{1}{n}\Sigma x_i$. This means that the mean, written as $\bar{x}$, is equal to the total of the sample divided by the sample size, n. The same formula is used for a population mean; however, μ (pronounced "mu") is the symbol for a population mean.

A distribution may be heavily skewed in one direction. For example, suppose an average delivery time is 22 minutes. However, 90% of deliveries took less than 17 minutes, but one took more than 90 minutes due to a delay caused by an accident. The resulting mean could be misleading.

Example

A student in a Six Sigma class earned the following quiz scores: 78, 88, 98, 89, 83, 98, 96, 91, 91, 80, 91, and 74. What is the mean?

Procedure

Step 1: Add all the individual quiz scores and then divide by the total number of quiz scores.

$$\frac{78 + 88 + 98 + 89 + 83 + 98 + 96 + 91 + 91 + 80 + 91 + 74}{12} = \frac{1057}{12}$$

Conclusion

The mean is 88.08.

Minitab Instructions

Step 1: Enter the data into a Minitab worksheet.

Step 2: Go to "Stat > Basic Statistics > Display Descriptive Statistics…"

Step 3: Click in the "Variables" box, then choose the appropriate column and click Select.

Step 4: (Optional) Clicking on "Statistics…" opens a box where additional descriptive statistics can be selected.

Step 5: (Optional) Clicking on "Graphs" opens a window where graphs can be selected.

Step 6: Click OK and the results will appear in the session window.

Minitab Output

Descriptive Statistics: C1

Variable	N	N*	Mean	SE Mean	StDev	Minimum	Q1	Median	Q3	Maximum
C1	12	0	88.08	2.26	7.84	74.00	80.75	90.00	94.75	98.00

Interpretation of Minitab Output

Minitab has determined the mean is 88.08.

Mode

The mode is the number that occurs most frequently in a data set. There can be multiple modes if more than one number is tied for most frequently occurring. The mode contains less information than the mean and the median.

Example

Find the mode of the data set from the previous example. The data set was: 78, 88, 98, 89, 83, 98, 96, 91, 91, 80, 91, and 74.

Procedure

Step 1: Determine which number occurs the most. There will be two modes if several numbers tie for the most frequently occurring number.

Conclusion

The number 91 occurs most often and therefore is the mode.

Minitab Instructions

Step 1: Enter the data into a Minitab worksheet.

Step 2: Go to "Stat > Basic Statistics > Display Descriptive Statistics…"

Step 3: Click in the "Variables" box, then choose the appropriate column and click Select.

Step 4: Clicking on "Statistics…"opens a box where additional descriptive statistics can be selected. Use this to select the mode.

Step 5: (Optional) Clicking on "Graphs" opens a window where graphs can be selected.

Step 6: Click OK and the results will appear in the session window.

Minitab Output

Descriptive Statistics: C1

Variable	N	N*	Mean	SE Mean	StDev	Minimum	Q1	Median	Q3	Maximum	Mode	Mode
C1	12	0	88.08	2.26	7.84	74.00	80.75	90.00	94.75	98.00	91	3

Interpretation of Minitab Output

Minitab has determined the mode is 91.

Median

The median of a data set is the 50th percentile of the distribution of the data. It is found by dividing the data set into two halves. The median of a data set with an odd-numbered sample size is the number in the middle of the ordered data set. The median of a data set with an even-numbered sample size is the average of the two numbers in the middle of the ordered data set.

An average delivery time of 22 minutes could be misleading if one delivery took more than 90 minutes due to an accident and all others took less than 17 minutes. In this case, the median delivery time of 14 would provide more useful information than the mean alone.

Example

Determine the median using the previous data set: 78, 88, 98, 89, 83, 98, 96, 91, 91, 80, 91, and 74.

Procedure

Step 1: Arrange the numbers in order from least to greatest and look for the number in the middle. Divide the two middle numbers by two if the data set has an even number of items. The order is: 74, 78, 80, 83, 88, 89, 91, 91, 91, 96, 98, 98.

Step 2: The data set has an even number of items and 89 and 91 are in the middle, so the median is equal to 89 + 91 / 2 = 90.

Conclusion

The median is 90.

Minitab Instructions

Step 1: Enter the data into a Minitab worksheet.

Step 2: Go to "Stat > Basic Statistics > Display Descriptive Statistics…"

Step 3: Click in the "Variables" box, then choose the appropriate column and click Select.

Step 4: (Optional) Clicking on "Statistics…" opens a box where additional descriptive statistics can be selected.

Step 5: (Optional) Clicking on "Graphs" opens a window where graphs can be selected.

Step 6: Click OK and the results will appear in the session window.

Minitab Output

Descriptive Statistics: C1

Variable	N	N*	Mean	SE Mean	StDev	Minimum	Q1	Median	Q3	Maximum	Mode	Mode
C1	12	0	88.08	2.26	7.84	74.00	80.75	90.00	94.75	98.00	91	3

Interpretation of Minitab Output

Minitab has determined the median is 90.00.

Measures of Variability

Measures of variability include the variance, standard deviation, range, and interquartile range (Weimar 1993). Variance and standard deviation are used to measure the variability or spread of data. The two values are closely related; the standard deviation is the square root of the variance. The variance is the summation of the average squared deviation of each number from the mean divided by the number of items in the population, or 1 minus the number of items in the sample or population. The interquartile range is the difference between the third quartile and the first quartile. The range is simply the difference between the highest and lowest values.

The Empirical Rule

The empirical rule can be used when data have an approximately normal distribution. It is used to determine the percentage of data values that are within a given number of standard deviations of the mean (Anderson, Sweeney, and Williams 2011). A standard normal distribution following the empirical rule is displayed in Figure 2.7.

Empirical rule for populations (or bell-shaped distributions):

1. Approximately 68% of all values are between one standard deviation below the mean and one standard deviation above the mean

2. Approximately 95% of all values are between two standard deviations below the mean and two standard deviations above the mean

3. Approximately 99.7% of all values are between three standard deviations below the mean and three standard deviations above the mean

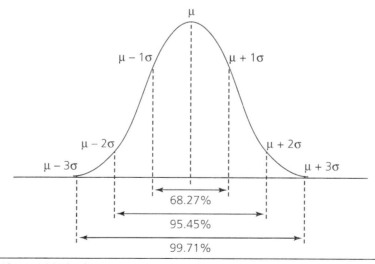

Figure 2.7   The empirical rule.

Example

A manufacturing engineer working on a Six Sigma project has a large sample of steel bars that have a mean Brinell hardness of 1620 HB and a standard deviation of 45. The sample population has a normal distribution, so the engineer can conclude that approximately 68% of all samples have a hardness of between 1575 HB and 1665 HB.

Chebyshev's Inequality

Chebyshev's inequality can be used when a population distribution is unknown. The great advantage of Chebyshev's inequality is that it is very general compared to the empirical rule; however, the empirical rule gives a better estimate when the distribution is known to be normal.

   Chebyshev's inequality (without assumptions regarding population distribution):

1. At least 75% of all values are between two standard deviations below the mean and two standard deviations above the mean

2. At least 89% of all values are between three standard deviations below the mean and three standard deviations above the mean

3. Generally, for any number $k > 1$, at least $(1 - 1 / k^2)$ of the population values will be within plus or minus k standard deviations of the mean

Example

A chemical process has a mean of 1200, a standard deviation of 150, and an unknown distribution. Using Chebyshev's inequality, a chemist could conclude that 75% of the process output is between 900 and 1500.

Range

The range is the simplest measure of the dispersion of a data set. The range is the difference between the highest and lowest values in a data set. The usefulness of the range may be limited due to its sensitivity to outliers in the data set.

Example

Data set: 132, 134, 122, 122, 114, 128, 136, 127, 122, 112, and 134.

Procedure

Step 1: Identify the highest number in the data set. It is 136 in this example.

Step 2: Identify the lowest number in the data set. It is 112 in this example.

Step 3: Subtract the lowest number from the highest number. The answer is the range.

Highest number – lowest number = 136 – 112 = 24

Conclusion

The range is 24.

Minitab Instructions

Step 1: Enter the data into a Minitab worksheet.

Step 2: Go to "Stat > Basic Statistics > Display Descriptive Statistics…"

Step 3: Click in the "Variables" box, then choose the appropriate column and click Select.

Step 4: (Optional) Clicking on "Statistics…"opens a box where additional descriptive statistics can be selected.

Step 5: (Optional) Clicking on "Graphs" opens a window where graphs can be selected.

Step 6: Click OK and the results will appear in the session window.

Minitab Output

Descriptive Statistics: C1

Variable	N	N*	Mean	SE Mean	StDev	Minimum	Q1	Median	Q3	Maximum
C1	11	0	125.73	2.44	8.10	112.00	122.00	127.00	134.00	136.00

Interpretation of Minitab Output

The range can be calculated by subtracting the minimum of 112 from the maximum, which is 136. The resulting range is 24.

Variance and Standard Deviation of a Population

The standard deviation of a population is the square root of the variance. A higher standard deviation indicates that the data are more widely scattered as compared to a lower standard deviation. The standard deviation is preferred over the variance because it is reported in the same units as the original measurements. The standard deviation of a population is identified by "σ." The standard deviation of a population is calculated with the formula $\sigma = \sqrt{\sigma^2}$; however, the variance must first be determined using this formula:

$$\sigma^2 = \frac{\Sigma(x_i - \mu)^2}{n}$$

Example

A production process with a standard normal deviation produces a batch of 5 metal rings. An engineer measures the inside diameter of all 5 parts and finds the following results in millimeters: 12.7, 12.4, 12.7, 12.6, and 12.5. Find the standard deviation.

Procedure

Step 1: Determine the average (μ) for the population: 12.7 + 12.4 + 12.7 + 12.6 + 12.5 = 62.9 / 5 = 12.58.

Step 2: Create a table with the following headings: x_i, μ, $x_i - \mu$ and $(x_i - \mu)^2$.

Step 3: Enter the individual data points in the x_i column.

Step 4: Enter the average for the data set in the µ column.

Step 5: Subtract the average (µ) from each data point (x_i) and enter the results directly to the right in the $x_i - \mu$ column.

Step 6: Calculate the values and then the total for the $(x_i - \mu)^2$ column and enter the result at the bottom of the table.

x_i	µ	$x_i - \mu$	$(x_i - \mu)^2$
12.7	12.58	0.12	0.0144
12.4	12.58	−0.18	0.0324
12.7	12.58	0.12	0.0144
12.6	12.58	0.02	0.0004
12.5	12.58	−0.08	0.0064
		Total:	0.068

Step 7: To determine the variance (σ^2), multiply the total for the $(x_i - \mu)^2$ column by 1 divided by the size of the population:

$$0.068 \times 1 / 5 = 0.068 \times 0.2 = 0.0136$$

Step 8: To determine the standard deviation (σ), find the square root of the variance (σ^2):

$$\sqrt{0.0136} = 0.117$$

Conclusion

The standard deviation is 0.117 millimeters. *Note:* Minitab only provides the standard deviation for a sample.

Variance and Standard Deviation of a Sample

The standard deviation of a sample is the square root of the sample variance. A higher standard deviation indicates that there is more variation scattered above and below the mean compared to a lower standard deviation. The standard deviation of a sample is identified by "S." The standard deviation of a sample is calculated with the following formula:

$$S = \sqrt{s^2}$$

However, the variance must first be found using this formula:

$$S^2 = \frac{\Sigma(x_i - \overline{x})^2}{n - 1}$$

Example

A large quantity of bulk material has arrived in drums. A quality technician takes samples of material from 5 of the drums and weighs the samples. The results are given in kilograms: 84, 86, 84, 85, and 83. Find the standard deviation of the sample.

Procedure

Step 1: Determine the average ($\bar{x}$) for the sample. The average is 84.4.

Step 2: Create a table with the following headings: x_i, $\bar{x}$, $x_i - \bar{x}$, and $(x_i - \bar{x})^2$.

Step 3: Enter the individual data points in the x_i column.

Step 4: Enter the average for the data set in the $\bar{x}$ column.

Step 5: Subtract the average ($\bar{x}$) from each data point (x_i) and enter the results directly to the right in the $x_i - \bar{x}$ column.

Step 6: Calculate the values and then the total for the $(x_i - \bar{x})^2$ column and enter the result at the bottom of the table.

x_i	$\bar{x}$	$x_i - \bar{x}$	$(x_i - \bar{x})^2$
84	84.4	−0.4	0.16
86	84.4	1.6	2.56
84	84.4	−0.4	0.16
85	84.4	0.6	0.36
83	84.4	−1.4	1.96
		Total:	5.2

Step 7: To determine the variance (S^2), divide the total for the $(x_i - \bar{x})^2$ column by the degrees of freedom, which is the size of the sample minus 1 ($n - 1$).

$$5.2 / (5 - 1) = 1.3$$

Step 8: To determine the standard deviation (S), find the square root of the variance (S^2).

$$\sqrt{1.3} = 1.14$$

Conclusion

The standard deviation is 1.14 kilograms.

Minitab Instructions

Step 1: Enter the data into a Minitab worksheet.

Step 2: Go to "Stat > Basic Statistics > Display Descriptive Statistics…"

Step 3: Click in the "Variables" box, then choose the appropriate column and click Select.

Step 4: (Optional) Clicking on "Statistics..." opens a box where additional descriptive statistics can be selected.

Step 5: (Optional) Clicking on "Graphs" opens a window where graphs can be selected.

Step 6: Click OK and the results will appear in the session window.

Minitab Output

Descriptive Statistics: C1

Variable	N	N*	Mean	SE Mean	StDev	Minimum	Q1	Median	Q3	Maximum
C1	5	0	84.400	0.510	1.140	83.000	83.500	84.000	85.500	86.000

Interpretation of Minitab Output

The Minitab result shows a standard deviation of 1.14.

Probability Plots

Many statistical tests and some types of statistical process control (SPC) require data to be normally distributed. A hypothesis test that has been performed using data that are not normally distributed could yield inaccurate results that lead a Six Sigma team to the wrong conclusion. An assessment of normality should be performed anytime data are required to be normally distributed.

Normality can be assessed by creating a normal probability plot. The cumulative frequency of each data point is plotted; together, the points should form an approximately straight line if the data are normally distributed. A probability plot can be constructed on probability paper by plotting the order values (j) against the observed cumulative frequency ($j - 0.5 / n$); however, this method can be subjective (Montgomery, Runger, and Hubele 2001). Ideally, normality should be assessed with the help of a statistical program.

Example

A Six Sigma Master Black Belt would like to perform a statistical hypothesis test on the following sample data: 48.34, 46.76, 47.74, 47.33, 48.43, 47.87, 47.95, and 47.11. The hypothesis requires data that follow a normal distribution, so the Six Sigma Master Black Belt must construct a probability plot to evaluate the normality of the data.

Procedure

Step 1: Determine the observed cumulative frequency by creating a table with the individual data points proceeding from lowest to highest.

Step 2: Calculate $(j - 0.5 / n)$ for each data point.

j	x_j	$(j - 0.5)/8$
1	46.76	.06
2	47.11	.19
3	47.33	.31
4	47.74	.44
5	47.87	.56
6	47.95	.69
7	48.34	.81
8	48.43	.94

Step 3: Plot the data and draw a straight line between the data points.

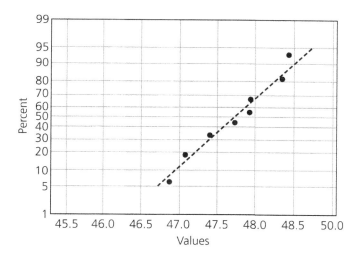

Conclusion

The data appear to be normally distributed.

Minitab Instructions

Step 1: Go to "Stat > Graph > Probability Plot…"

Step 2: Select "Single" and click OK.

Step 3: Select the column containing the data, then click OK.

Minitab Output

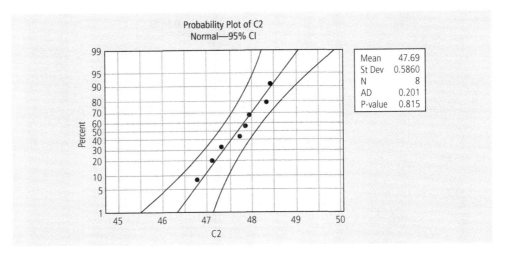

Probability Plot of C2
Normal—95% CI

Mean	47.69
St Dev	0.5860
N	8
AD	0.201
P-value	0.815

Interpretation of Minitab Output

The probability plot can be seen in the Minitab-generated graph. Minitab also gives results for the Anderson-Darling Test of Normality and a *P* value. The *P* value of 0.815 is greater than 0.05, so we fail to reject the hypothesis that the distribution is equal to a normal distribution.

Box Plots

Box plots, also known as box-and-whisker plots, are graphical representations of a data set. A box plot identifies the median and includes a box containing 50% of the data set. The whiskers on a box plot each contain 25% of the data, excluding outliers. Box plots can be used to quickly compare multiple data sets to gain a better understanding of the data.

Example

A Six Sigma team member collected the following data to create a box plot: 106.2, 104.7, 104.5, 103.1, 104.7, 105.4, 105.0, 105.8, 104.8, 104.6, 104.2, 107.4, 105.0, and 104.6.

Procedure

Step 1: Place the data set in order from lowest to highest: 103.1, 104.2, 104.5, 104.6, 104.6, 104.7, 104.7, 104.8, 105.0, 105.0. 105.4, 105.8, 106.2, 107.4.

Step 2: Determine the median by finding the value in the middle of the data set (or by averaging the two values in the middle if using an even data set). The median is the average of 104.7 and 104.8; therefore, the median is 104.75.

Step 3: Find the first quartile (Q1). This is the middle point from the lowest number to the number below the median. This is from 103.1 to 104.7; the midpoint is 104.5.

Step 4: Find the third quartile (Q3). This is the middle point from the number above the median to the highest number. This is from 104.8 to 107.4; the midpoint is 105.4.

Step 5: Find the lower end of the lower whisker. This is equal to Q1 − 1.5(Q3 − Q1); therefore, 104.5 − 1.5(105.4 − 104.5) = 103.15.

Step 6: Find the upper end of the upper whisker. This is equal to Q3 + 1.5(Q3 − Q1); therefore, 105.4 + 1.5(105.4 − 104.5) = 106.75.

Step 7: Create a number line that encompasses the data set.

Step 8: Use vertical lines to represent Q1, the median, and Q3. Connect them with horizontal lines. Draw a horizontal line from Q1 to the end of the lower whisker and a horizontal line from Q3 to the end of the upper whisker. Use an asterisk to identify any point beyond the whiskers as an outlier.

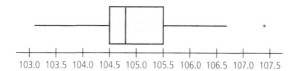

103.0 103.5 104.0 104.5 105.0 105.5 106.0 106.5 107.0 107.5

Conclusion

There is one outlier present in the data set.

Minitab Instructions

Step 1: Go to "Stat > Graph > Boxplot..."

Step 2: Select "Simple" and click OK.

Step 3: Use the column containing the data for Graph variables, then click OK.

Minitab Output

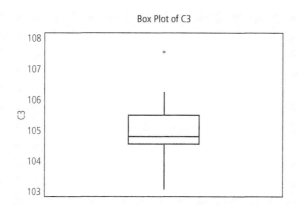

Box Plot of C3

Interpretation of Minitab Output

Minitab generated a box plot of the data set with one outlier.

Measures of Shape

The shape of a histogram is determined by the relative position of the mean, mode, and median (Weimar 1993). A rough idea of the shape of a distribution can be seen by observing a histogram; however, the exact shape can be calculated by using the skew and kurtosis of the distribution.

Skew

Skewness ranges from –3 to 3. Perfectly symmetrical data have a skewness equal to 0 because the mean is equal to the median. The data are skewed to the right when the mean is larger than the median. In such situations, values above the mean occur less frequently than values below the mean, and the tail of the distribution is toward the right side. The formula for skewness is Pearson's coefficient of skewness:

$$Sk = \frac{3(\bar{x} - Md)}{s}$$

with Md representing the median.

Example

A manufacturing process has mean of 13.643, a standard deviation of 1.692, and a median of 14.000.

Procedure

Step 1: Subtract the median from the mean: 13.643 – 14.000 = –0.357.

Step 2: Multiply the result by 3: 3 × –0.357 = –1.071.

Step 3: Divide the result by the standard deviation: –1.071 / 1.692 = –0.633.

Conclusion

The process is moderately skewed to the left.

Minitab Instructions

Note: These instructions apply when the raw data are available as a column.

Step 1: Enter the data into a Minitab worksheet.

Step 2: Go to "Stat > Basic Statistics > Display Descriptive Statistics…"

Step 3: Click in the "Variables" box, then choose the appropriate column and click Select.

Step 4: Clicking on "Statistics…" opens a box where additional descriptive statistics can be selected. Use this to select Skew.

Step 5: (Optional) Clicking on "Graphs" opens a window where graphs can be selected.

Step 6: Click OK and the results will appear in the session window.

Minitab Output

Descriptive Statistics: C1

Variable	Mean	Median	Skewness
C1	13.643	14.000	−0.68

Interpretation of Minitab Output

Minitab indicates the skewness is −0.68, which is slightly different from the result from Pearson's coefficient of skewness. Statistical software packages use a more complicated formula that uses the summation of the individual Z scores for each value. For Minitab, this measure of skew is called the adjusted Fisher-Pearson standardized moment coefficient (Sk').

Kurtosis

Kurtosis measures peakedness and how heavy-tailed a distribution is. This can be understood as the height and sharpness of a distribution. A higher kurtosis value indicates a sharper peak and a lower kurtosis value indicates a smaller, less sharp peak. A normal distribution has a kurtosis of 3. A distribution with a kurtosis approximately equal to 3 is called mesokurtic. A distribution with a kurtosis less than 3 is called platykurtic, and a distribution with a kurtosis greater than 3 is called leptokurtic. The formula for calculating kurtosis is:

$$\frac{n(n + 1)}{(n - 1)(n - 2)(n - 3)} \ \Sigma[(x_i - \overline{x})/s]^4 - \frac{3(n - 1)^2}{(n - 2)(n - 3)}$$

Example

Eleven samples were checked and had the following results: 48, 51, 52, 49, 46, 53, 56, 52, 51, 48, and 56.

Procedure

Step 1: Determine the number of observations. This is 11; therefore, $n = 11$.

Step 2: Determine the mean. The mean is 51.09; therefore $\overline{x} = 51.09$.

Step 3: Determine the standard deviation. The standard deviation is 3.41; therefore, $\sigma = 3.208$.

Step 4: Determine $\sum[(x_i - \bar{x})/s]^4$ by creating a table as depicted below.

x_i	$\bar{x}$	$x_i - \bar{x}$	s	$(x_i - \bar{x})/S$	$((x_i - \bar{x})/S)^4$
48	51.09	−3.09	3.208	−0.962616822	0.8607885
51	51.09	−0.09	3.208	−0.28037383	0.0000006
52	51.09	0.91	3.208	0.283665835	0.0064748
49	51.09	−2.09	3.208	−0.963216958	0.1801556
46	51.09	−5.09	3.208	−1.586658354	6.3377294
53	51.09	1.91	3.208	0.595386534	0.1256597
56	51.09	4.91	3.208	1.530548628	5.4876769
52	51.09	0.91	3.208	0.283665835	0.0064748
51	51.09	−0.09	3.208	−0.028054863	0.0000006
48	51.09	−3.09	3.208	−0.963216958	0.8607885
56	51.09	4.91	3.208	1.530548628	5.4876769
				Total:	19.353

Step 5: Plug the answer from Step 4 into the formula and fill in the rest of the formula.

$$\frac{n(n + 1)}{(n - 1)(n - 2)(n - 3)} \sum[(x_i - \bar{x})/s]^4 - \frac{3(n - 1)^2}{(n - 2)(n - 3)} =$$

$$\frac{11(11 + 1)}{(11 - 1)(11 - 2)(11 - 3)} \, 19.353 - \frac{3(11 - 1)^2}{(11 - 2)(11 - 3)} =$$

$$(0.183)(19.353) - 4.167 = -0.625$$

Conclusion

The kurtosis is −0.625. This is much less than 3, so the distribution is platykurtic.

Minitab Instructions

Step 1: Enter the data into a Minitab worksheet.

Step 2: Go to "Stat > Basic Statistics > Display Descriptive Statistics…"

Step 3: Click in the "Variables" box, then choose the appropriate column and click Select.

Step 4: Clicking on "Statistics…" opens a box where additional descriptive statistics can be selected. Use this to select Kurtosis.

Step 5: (Optional) Clicking on "Graphs" opens a window where graphs can be selected.

Step 6: Click OK and the results will appear in the session window.

Minitab Output

Descriptive Statistics: C1

Variable	N	N*	Mean	SE Mean	StDev	Minimum	Q1	Median	Q3	Maximum	Kurtosis
C1	11	0	51.091	0.967	3.208	46.000	48.000	51.000	53.000	56.000	−0.62

Interpretation of Minitab Output

Minitab has determined that the kurtosis is −0.62.

CAPABILITY AND PERFORMANCE STUDIES

Process capability is measured by Cp and Cpk. The Cp is a capability index; it compares the variability in a process to the maximum variability permitted by the tolerance. A process capability index quantifies the ability of a process to conform to specifications (Burrill and Ledolter 1999). The Cpk is like the Cp but it takes the process location into account. The Cp only tells us how the data are spread relative to the tolerance limits; the Cpk considers where the process is centered. It is possible for a process with a very good Cp to be completely outside one of the tolerances. A Cp and Cpk is used when a process is in control and it uses the pooled standard deviation.

Process performance is measured with Pp and Ppk. The Pp is much like the Cp, but it is used when a process is not in statistical control. Similarly, the Ppk is like the Cpk but is only used when a process is not in statistical control and it uses the total standard deviation. Both Cp and Pp tell us the potential of a process; they are the ratio of the width of the upper specification limit (USL) and the lower specification limit (LSL) relative to 6 times the process variation:

$$Cp = \frac{USL - LSL}{6\sigma_P} \qquad Pp = \frac{USL - LSL}{6\sigma_T}$$

This does not tell us how the process is actually operating, because the positioning of the process is not accounted for. To determine the actual performance of the process, Cpk or Ppk is used. The calculations for Cpk and Ppk need to be performed twice—once for each specification limit. The lower Cpk or Ppk is the true value for the process. The pooled standard deviation (σ_P) is taken over a short time and only has common cause variation within the subgroup. The total standard deviation

(σ_T) is taken over a longer period of time and includes both common cause and special cause variation.

$$\text{Cpk} = \frac{\bar{x} - \text{LSL}}{3\sigma_P}, \frac{\text{USL} - \bar{x}}{3\sigma_P} \quad \text{Ppk} = \frac{\bar{x} - \text{LSL}}{3\sigma_T}, \frac{\text{USL} - \bar{x}}{3\sigma_T}$$

A process can be considered capable if the Cp is equal to or greater than 1; a Cp of 2 indicates the process is capable of performing at a Six Sigma level. The Cpk must also be considered, because Cp does not account for the positioning of the process relative to the specification limits. Cpk levels of 1.33 or 1.67 are typically used as targets (Benbow and Kubiak 2009).

Cp and Cpk

Using Cp and Cpk tells us what a process is capable of.

$$\text{Cp} = \frac{\text{USL} - \text{LSL}}{6\sigma_P}$$

$$\text{Cpk} = \frac{\bar{x} - \text{LSL}}{3\sigma_P}, \frac{\text{USL} - \bar{x}}{3\sigma_P}$$

Example

A Six Sigma Black Belt collects 4 subgroups of sample size 4 from a machining process. The specification is 32.4 with a tolerance of +/– 0.4. The results are in the table below.

Subgroup 1	Subgroup 2	Subgroup 3	Subgroup 4
32.37	32.48	32.26	32.47
32.45	32.51	32.41	32.33
32.40	32.46	32.32	32.45
32.37	32.43	32.34	32.33

Procedure

Step 1: Determine the pooled process standard deviation σ_P; if that is unknown, calculate the sample standard deviation. The sample standard deviation is 0.07.

Step 2: Calculate $\text{Cp} = \dfrac{\text{USL} - \text{LSL}}{6\sigma} = (32.8 - 32.0) / (6)(0.07) = 1.90$.

Step 3: Find the mean of the samples. The mean is 32.4.

Step 4: Calculate the Cpk for the USL:

$$Cpk = \frac{USL - \bar{x}}{3\sigma} = (32.8 - 32.4)/(3)(0.07) = 1.90$$

Step 5: Calculate the Cpk for the LSL:

$$Cpk = \frac{\bar{x} - LSL}{3\sigma} = (32.4 - 32.0)/(3)(0.07) = 1.90$$

Conclusion

The Cp and Cpk are both equal to 1.90.

Minitab Instructions

Step 1: Go to "Stat > Quality tools > Capability analysis > Normal…"

Step 2: Select the column containing the data; use the column for the subgroup size or enter a subgroup size.

Step 3: Enter the upper and lower specification limits and click OK.

Minitab Output

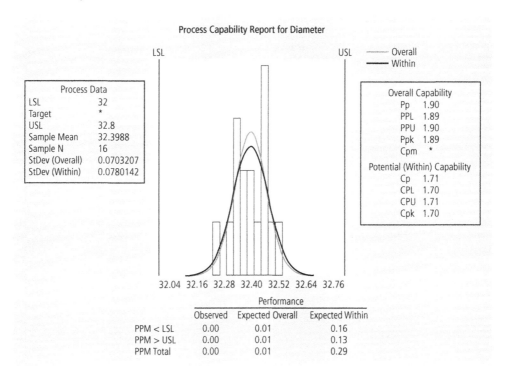

Process Capability Report for Diameter

Process Data	
LSL	32
Target	*
USL	32.8
Sample Mean	32.3988
Sample N	16
StDev (Overall)	0.0703207
StDev (Within)	0.0780142

Overall Capability
Pp 1.90
PPL 1.89
PPU 1.90
Ppk 1.89
Cpm *

Potential (Within) Capability
Cp 1.71
CPL 1.70
CPU 1.71
Cpk 1.70

32.04 32.16 32.28 32.40 32.52 32.64 32.76

Performance			
	Observed	Expected Overall	Expected Within
PPM < LSL	0.00	0.01	0.16
PPM > USL	0.00	0.01	0.13
PPM Total	0.00	0.01	0.29

Interpretation of Minitab Output

Minitab has provided the Cp and Cpk. Minitab has also provided predictions for the process performance. Note: Our results differ from the Minitab results because we used the sample standard deviation and Minitab used the pooled standard deviation.

Pp and Ppk

The Pp and Ppk should be used to determine what the process is doing long term.

$$Pp = \frac{USL - LSL}{6\sigma_T}$$

$$Ppk = \frac{\bar{x} - LSL}{3\sigma_T}, \frac{USL - \bar{x}}{3\sigma_T}$$

Example

The specification for the length of a tube is 67.6 with a tolerance of +/– 0.6. A Six Sigma Green Belt collects a sample of 8 parts with the following results: 67.73, 67.35, 67.63, 67.79, 67.50, 68.06, 67.55, and 67.46.

Procedure

Step 1: Determine the standard deviation σ_T. The standard deviation is 0.223.

Step 2: Find $Pp = \dfrac{USL - LSL}{6\sigma_T}$ = (68.2 – 67.0) / (6)(0.223) = 0.90.

Step 3: Calculate $Ppk = \dfrac{USL - \bar{x}}{3\sigma_T}$ = (68.2 – 67.63) / (3)(0.223) = 0.85.

Step 4: Calculate $Ppk = \dfrac{\bar{x} - LSL}{3\sigma_T}$ = (67.63 – 67.0) / (3)(0.223) = 0.94.

Conclusion

The Pp is 0.90. The Ppk for the USL is 0.85 and the Ppk for the LSL is 0.94; we use the Ppk of 0.85 as the value for the process.

Minitab Instructions

Step 1: Go to "Stat > Quality tools > Capability analysis > Normal..."

Step 2: Select the column containing the data; use the column for the subgroup size or enter a subgroup size.

Step 3: Enter the upper and lower specification limits and click OK.

Minitab Output

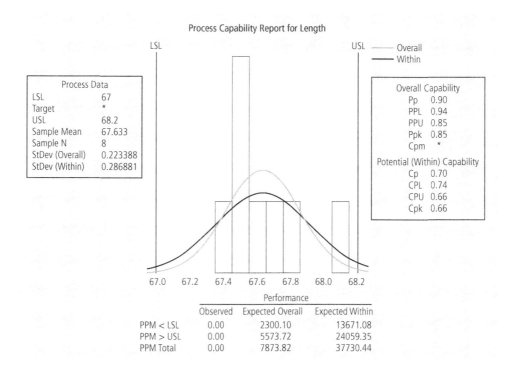

Process Capability Report for Length

Process Data	
LSL	67
Target	*
USL	68.2
Sample Mean	67.633
Sample N	8
StDev (Overall)	0.223388
StDev (Within)	0.286881

Overall Capability	
Pp	0.90
PPL	0.94
PPU	0.85
Ppk	0.85
Cpm	*

Potential (Within) Capability	
Cp	0.70
CPL	0.74
CPU	0.66
Cpk	0.66

Performance

	Observed	Expected Overall	Expected Within
PPM < LSL	0.00	2300.10	13671.08
PPM > USL	0.00	5573.72	24059.35
PPM Total	0.00	7873.82	37730.44

Interpretation of Minitab Output

Minitab has provided the Pp and Ppk for the USL. Minitab has also provided predictions for the process performance.

MEASUREMENT SYSTEM ANALYSIS

Measurement system analysis (MSA) is performed to analyze the way in which measurement data are acquired; it assesses whether variability in measurement data is caused by the measuring system, which includes the measuring device and the operator of the device. Suppose a metal block is measured three times at the exact same location and each measurement yields a different result; this would indicate there is variability resulting from the measurement system.

Important concepts to consider when assessing a measurement system are true value, discrimination, number of distinct categories, accuracy (bias), precision, stability, and linearity. The true value is the actual value of the part being measured; this value can never truly be known, but it should be identified as precisely as possible, taking into consideration the economics of gaining a more exact value (Chrysler, Ford, and General Motors 2010). For example, knowing the exact length of a brick is not as important as knowing the diameter of an automotive engine's piston, so lower-cost measuring devices can be used for the brick.

Discrimination, also known as resolution, is the ability of a measuring device to read changes in a reference value. A measuring device should be capable of measuring a minimum of one tenth of a part's specification range. Problems with discrimination can be identified by the number of distinct categories the measuring device can detect.

The number of distinct categories is the number of separate categories into which the measurement data can be separated. A measuring device that can only detect one distinct category is only useful for determining whether a part is in specification or out of specification. An example of a measuring device with only one distinct category is calipers that consistently give the same reading when measuring the thickness of multiple pieces of sheet metal. The calipers are unable to detect minor variations in the thickness. Two to four distinct categories can only provide rough estimates of the true value; such measuring devices should not be used for SPC charts for variable data. A measuring device should have five distinct categories (Chrysler, Ford, and General Motors 2010).

The difference between the average of measured values and the true value is called bias. Bias is also referred to as accuracy; however, the term "accuracy" has other meanings and should be avoided to prevent confusion. Precision, also known as repeatability, is the variation in repeated measurements of the same part by the same person. Stability refers to changes in the measurements of the same characteristic over time (Gryna 2001). Figure 2.8 depicts the difference between precision and accuracy.

Changes in bias over a range of measurements are due to linearity (Borror 2009). This means that bias changes as the size of the measured part changes. For example, a micrometer has a linearity problem if it has little bias at the bottom of its measuring range but loses precision toward the top of its measuring range.

A gage repeatability and reproducibility (gage R&R) analysis is performed to assess the repeatability and reproducibility of a measurement system. Repeatability is variability when the same measurements are made under the same conditions. Reproducibility is the variation resulting when the same measurements are made under the same conditions by different operators (George et al. 2005). A gage R&R can be performed using an analysis of variance (ANOVA) table as shown in Table 2.3. (See Chapter 3 for details on creating an ANOVA table.)

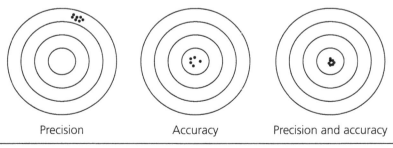

Precision   Accuracy   Precision and accuracy

Figure 2.8   Precision and accuracy.

Table 2.3 ANOVA table.

Source	df	SS	MS
Part	p − 1	SS_P	MS_P
Operator	o − 1	SS_O	MS_O
Part X operator	(p − 1)(o − 1)	SS_{OP}	MS_{OP}
Error	op(n − 1)	SS_E	MS_E
Total	ppn − 1	SS_T	

Reproduced from Connie M. Borror, ed., *The Certified Quality Engineer Handbook* (3rd ed.). Milwaukee, WI: ASQ Quality Press, 2009. Reprinted with permission.

Variance due to parts and operators:

$$\hat{\sigma}_O^2 = \frac{MS_O - MS_{OP}}{pr}$$

where p = number of parts and r = number of replicates.
Variance due to operators:

$$\hat{\sigma}_{PO}^2 = \frac{MS_{OP} - MS_E}{r}$$

where r = number of replicates.
Variance due to parts:

$$\hat{\sigma}_P^2 = \frac{MS_P - MS_{OP}}{or}$$

where o = number of operators and r = number of replicates.

$\hat{\sigma}_{\text{Reproducibility}}^2$: $\hat{\sigma}_{PO}^2 + \hat{\sigma}_O^2$

$\hat{\sigma}_{\text{Repeatability}}^2$: Error $\hat{\sigma}_e^2 = MS_E$

Total gage R&R: $\hat{\sigma}_{\text{Measurement error}}^2 = \hat{\sigma}_{\text{Repeatability}}^2 = \hat{\sigma}_{\text{Reproducibility}}^2$

$$\text{Number of distinct categories} = 1.41 \frac{\text{Variance due to parts } \hat{\sigma}_P^2}{\text{Total gage R\&R } \hat{\sigma}_{\text{Measurement error}}^2}$$

The Automotive Industry Action Group considers any gage with more than 30% variation to be unacceptable. Less than 10% variation is considered acceptable. Between 10% and 30% variation may be acceptable, but improvements should be considered depending on the use of the measurements and the cost of improvements (Chrysler, Ford, and General Motors 2010).

Example

A Six Sigma Master Black Belt performed a gage R&R with 3 operators, 10 parts, and 3 replicates. The measurement results are depicted below.

Operator	Trial	Part number									
		1	**2**	**3**	**4**	**5**	**6**	**7**	**8**	**9**	**10**
1	1	83.9	83.8	84.1	84.8	85.5	83.8	84.1	84.0	84.3	85.1
	2	84.3	83.4	84.5	84.3	85.2	84.1	84.4	83.7	84.4	85.3
	3	84.1	83.8	84.2	84.7	85.4	84.2	84.3	84.1	84.1	85.3
2	1	84.4	83.9	84.6	85.4	85.8	84.6	84.5	84.7	84.6	85.6
	2	84.3	83.9	84.4	85.1	85.7	84.3	84.6	84.5	84.5	85.6
	3	84.2	83.7	84.2	85.1	85.4	84.1	84.7	84.4	84.4	85.7
3	1	84.3	84.1	84.5	84.9	85.5	84.5	84.1	83.7	84.6	84.8
	2	83.9	83.7	84.0	84.7	85.2	84.2	84.4	84.3	84.3	85.1
	3	84.2	83.6	83.9	84.3	85.1	83.8	83.4	84.5	84.1	85.0

Procedure

Step 1: Create an ANOVA table.

Source	df	SS	MS	F
Part	9	22.311	2.479	44.396
Operator	2	2.228	1.114	19.952
Part X operator	18	1.005	0.056	1.038
Repeatability	60	3.227	0.054	
Total	89	28.771		

Step 2: Use the F table in Appendix C to determine which components of the variance have a statistically significant difference. We fail to reject the null hypothesis for part X operator based on the F table.

Step 3: Find variance due to operators: $\hat{\sigma}_O^2 = \dfrac{MS_O - MS_{OP}}{pr} = \dfrac{1.114 - 0.056}{(10)(3)} = 0.035$.

Step 4: Find variance due to parts and operators: $\hat{\sigma}_{PO}^2 = \dfrac{MS_{OP} - MS_E}{r} = \dfrac{0.056 - 0.054}{3} = 0.0006$.

Step 5: Find variance due to parts: $\hat{\sigma}_P^2 = \dfrac{MS_P - MS_{OP}}{or} = \dfrac{2.479 - 0.056}{(3)(3)} = 0.269$.

Step 6: Find $\hat{\sigma}_{Reproducibility}^2$: $\hat{\sigma}_{PO}^2 + \hat{\sigma}_O^2 = 0.0006 + 0.035 = 0.036$.

Step 7: Find $\hat{\sigma}_{Repeatability}^2$: Error $\hat{\sigma}_e^2 = MS_E = 0.054$.

Step 8: Find total gage R&R: $\hat{\sigma}_{Measurement\ error}^2 = \hat{\sigma}_{Repeatability}^2 + \hat{\sigma}_{Reproducibility}^2 = 0.036 + 0.054 = 0.90$.

Step 9: Find the percent contribution of the total gage R&R by dividing total gage R&R by total gage R&R plus variance due to parts: 0.90 / (0.90 + 0.269) = 25.1%.

Step 10: Determine the number of distinct categories:

$$1.41 \dfrac{\text{Variance due to parts } \hat{\sigma}_P^2}{\text{Total gage R\&R } \hat{\sigma}_{Measurement\ error}^2} = 1.41(0.269/0.90) = 0.42.$$

Conclusion

The measurement system is capable with a total gage R&R of 25.1%; however, improvements should be considered based on the intended use and cost of improvements.

Minitab Instructions

Step 1: Go to "Stat > Quality tools > Gage study > Create gage R&R study worksheet…"

Step 2: The number of parts, operators, and replicates can be changed here. Change the number of replicates to 3 and click OK.

Step 3: Perform the gage R&R and use the Minitab gage R&R worksheet to collect the data.

Step 4: Go to "Stat > Quality tools > Gage study > Gage study R&R study (Crossed)…"

Step 5: Enter the columns containing part numbers, operators, and the measurement data into the appropriate fields and click OK.

Minitab Output

Gage R&R Study—ANOVA Method
Two-Way ANOVA Table With Interaction

Source	DF	SS	MS	F	P
Parts	9	22.3112	2.47902	44.3955	0.000
Operators	2	2.2282	1.11411	19.9520	0.000
Parts * Operators	18	1.0051	0.05584	1.0383	0.434
Repeatability	60	3.2267	0.05378		
Total	89	28.7712			

α to remove interaction term = 0.05

Two-Way ANOVA Table Without Interaction

Source	DF	SS	MS	F	P
Parts	9	22.3112	2.47902	45.6933	0.000
Operators	2	2.2282	1.11411	20.5353	0.000
Repeatability	78	4.2318	0.05425		
Total	89	28.7712			

Gage R&R

Source	VarComp	%Contribution (of VarComp)
Total Gage R&R	0.089582	24.95
Repeatability	0.054254	15.11
Reproducibility	0.035329	9.84
Operators	0.035329	9.84
Part-To-Part	0.269419	75.05
Total Variation	0.359001	100.00

Source	StdDev (SD)	Study Var (6 × SD)	%Study Var (%SV)
Total Gage R&R	0.299303	1.79582	49.95
Repeatability	0.232924	1.39754	38.87
Reproducibility	0.187959	1.12775	31.37
Operators	0.187959	1.12775	31.37
Part-To-Part	0.519056	3.11434	86.63
Total Variation	0.599167	3.59500	100.00

Number of Distinct Categories = 2

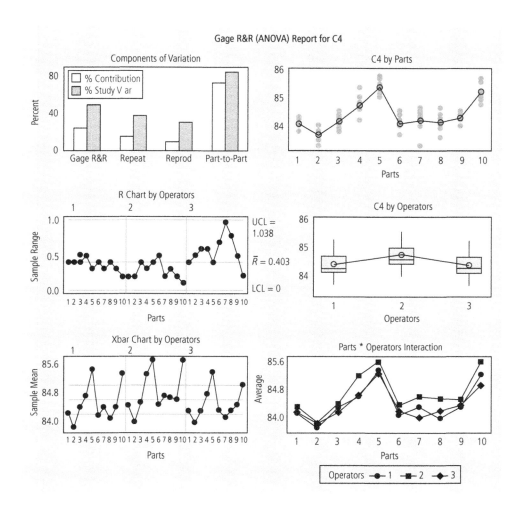

Interpretation of Minitab Output

Minitab shows the measurement system is capable with a total gage R&R of 24.95%; however, it also shows that the number of distinct categories is 2. The measuring device cannot give results that will be usable for SPC.

Chapter 3
Analyze

The data collected during the Measure phase are analyzed during the Analyze phase "to identify the causes of variation and process performance" (Gryna 2001). Hypothesis testing and regression analysis are often used during this phase of a Six Sigma project. The causes and relationships established during this phase are used during the Improvement phase.

HYPOTHESIS TESTING

One of the meanings of "hypothesis" in statistics is an assumption about a population parameter. Data from a sample are used to test hypotheses concerning the probable value of the true population parameter. Population parameters are designated by Greek letters and the corresponding statistics are designated with Roman letters.

The null hypothesis (H_0) is used as a contrast to what is being tested in the population and is stated in terms of $\leq$, $\geq$, or $=$ in regard to a specific value. The null hypothesis is compared to an alternative hypothesis (H_a). The null hypothesis is that there is no difference between the hypothesized results and the actual results. In hypothesis testing we often place whatever we are testing in the alternative hypothesis and then compare it against a null hypothesis. The null hypothesis can include $\leq$, $\geq$, or $=$. The alternative hypothesis can include $<$, $>$, or $\neq$.

A test can be one tailed or two tailed. A two-tailed test is used to test for a difference between the actual value and a hypothesized value. A one-tailed test is used when the hypothesized value is believed to deviate from the true value in a specific direction. The null hypothesis of a one-tailed test should include all options other than the one being evaluated in the alternative hypothesis (Witte 1993).

A two-tailed hypothesis could be:

$$H_0: \mu = 10$$

$$H_a: \mu \neq 10$$

A one-tailed test could be:

$$H_0: \mu \leq 10$$

$$H_a: \mu > 10$$

Or:

$$H_0: \mu \geq 10$$
$$H_a: \mu < 10$$

The alternative hypothesis is considered to be true if the null hypothesis has been rejected. Failure to reject the null hypothesis does not mean the null hypothesis is true; it only means there is not sufficient evidence to reject it. The null hypothesis is not accepted or confirmed; it is merely not rejected or rejected.

Kenney (1988) realizes that this way of testing a hypothesis may seem "illogical," so he compares hypothesis testing to a court trial. In his analogy, the null hypothesis is comparable to the assumption that a defendant is innocent, and the alternative hypothesis is comparable to the charge that the defendant is guilty. A judge or jury finding a defendant not guilty is not the same as them declaring the defendant to be innocent. Rather, there is insufficient evidence to conclude that the defendant is guilty; in the case of hypothesis testing, rejecting the null hypothesis is not the same as accepting the alternative hypothesis.

Error Types

There are two possible types of errors in hypothesis testing: the type I error, also known as an alpha (α) error, and the type II error, also known as a beta (β) error. A type I error is committed when the null hypothesis was rejected but is true. A type II error is committed when the alternative hypothesis is true but the null hypothesis was not rejected (see Table 3.1).

The power of a test refers to its ability to actually detect the phenomenon being tested. A hypothesis test with more power has a better chance of finding an actual difference in two unknown distributions than a test with a lower power (Hinton 2004). The power of a test is equal to $1 - \beta$.

A type I error is generally a more serious mistake than a type II error. It is usually better to incorrectly conclude that nothing has happened than to incorrectly conclude that there are effects present. For example, suppose a manufacturer is considering an investment to improve the accuracy of a machine. A two-tailed hypothesis could be:

$$H_0: \text{Accuracy of old machine} = \text{Accuracy of new machine } 98\%$$

$$H_a: \text{Accuracy of old machine} \neq \text{Accuracy of new machine } 98\%$$

Table 3.1   Error types.

Action	Null hypothesis (H_0) is true	Alternative hypothesis (H_a) is true
Reject null hypothesis (H_0):	Type I error (α)	Correct
Fail to reject null hypothesis (H_a):	Correct	Type II error (β)

Or a one-tailed hypothesis could be:

H_0: Accuracy of old machine $\geq$ Accuracy of new machine 98%

H_a: Accuracy of old machine $<$ Accuracy of new machine 98%

If we use a one-tailed test and the improved machine is not equal to or better than the old machine, but we fail to reject the null hypothesis, then a type II error has occurred. A type I error could result in missing an opportunity to improve performance, while a type II error could result in making an unnecessary change that could degrade performance.

A typical alpha level is 0.05, though other alpha levels such as 0.01 can also be used. Using an alpha of 0.05 means you're willing to accept a 5% chance of making a type I error. The rejection criteria for a hypothesis test should be determined before data are collected and the test is analyzed. Simmons, Nelson, and Simonsohn (2011) warn of what they call "researcher degrees of freedom" in publishing analyses that use hypothesis testing. Researchers can decide to collect additional data or exclude data from an analysis, or they may perform multiple analyses and only report the one that was statistically significant. Unfortunately, disregarding the insignificant results of previous trials and only accepting the significant results increases the chances of a false positive.

This is also relevant to industry. For example, suppose a machine produces a part that does not conform to the distribution of the other parts produced. This could be because of an outside influence that is not relevant to the experiment; hence, the outlier could potentially be discarded. Or, it could be an unknown factor that will always be present in the system, if only sporadically. The data should not be discarded in such a case. The researcher or investigator must make a judgment call in such a situation; however, an incorrect decision when discarding outliers can result in incorrect results. The source of the outlier should be thoroughly investigated before a choice is made. If an outlier is discarded, the test results should include a note indicating this.

Another potential problem can arise when an engineer performs hypothesis testing to evaluate the results of a process improvement and the results are not found to be statistically significant, so the engineer repeats the experiment several times using an alpha of 0.05. A significant result on the fifth try cannot be honestly reported as statistically significant at an alpha of 0.05; the significance level is only valid for the first trial, and the chance of a false positive increases with each new trial. A Bonferroni correction can be performed, but that is beyond the scope of this book. See Warner (2008) for details on the Bonferroni correction.

Sample size is also an important factor in hypothesis testing. There is a greater chance of failing to detect a statistically significant difference when a small sample size is used. Unfortunately, increasing the sample size may also increase costs due to the additional resources needed.

Five Steps for Hypothesis Testing

1. *Determine which test to use.* Select the appropriate test statistic for the test that will be performed, and decide if a one- or two-tailed test is required. If the test will be a one-tailed test, decide if the upper or lower tail will be used.

2. *Set the null hypothesis.* For a two-tailed hypothesis test, the null hypothesis is no difference. An example of a null hypothesis for a two-tailed test for one sample is, "The mean of the population is equal to 425," which would be written as "H_0: $\mu = 425$." For a one-tailed hypothesis, the null hypothesis is generally that the value being tested is less than or equal to the stated value (for an upper-tailed test) or greater than or equal to the stated value (for a lower-tailed test).

3. *Set the alternative hypothesis.* The null hypothesis is always evaluated against an alternative hypothesis. The alternative hypothesis is what we are interested in and is formed in such a way that it contrasts with the null hypothesis. The alternative hypothesis for the null hypothesis in the example above would be, "The mean of the population is not equal to 425," which would be written as "H_a: $\mu \neq 425$." For a one-tailed hypothesis, the alternative is generally that the value being evaluated is greater than (for an upper-tailed test) or less than (for a lower-tailed test) the value stated in the null hypothesis.

   The objective of the hypothesis test is not to prove the alternative hypothesis; rather, it is to determine whether there is sufficient evidence to reject the null hypothesis in favor of the alternative hypothesis.

4. *Set the alpha level and determine the rejection criteria.* There is no specific requirement for the alpha level; however, $\alpha = 0.05$ is commonly used and results in 95% confidence. An alpha of $\alpha = 0.01$ could be used for 99% confidence; however, the risk of a type II error increases as the chance of a type I error decreases.

5. *Reject or fail to reject the null hypothesis and form conclusions.* Perform the calculations, reject or fail to reject the null hypothesis, and then form conclusions. For example, if two processes are compared and one is found to be inferior to the other, the inferior process should be either improved or discontinued.

TESTS OF POPULATION MEANS

Z-Tests

A Z-test can be a one-tailed or two-tailed test. The rejection regions are at the two tails of the distribution in a two-tailed Z-test and at either the upper or lower tail of the distribution when using a one-tailed Z-test (Witte 1993). The rejection criteria of a test of means using a Z score are based on the alpha level selected. A commonly used alpha level is 0.05, which gives a confidence level of 95% and gives a critical Z score of 1.960 for a two-tailed test and 1.645 for a one-tailed test. Another possible alpha level is 0.01, which gives a confidence level of 99% and gives a critical Z score of 2.258 for a two-tailed test and 2.33 for a one-tailed test.

   The critical Z score defines the rejection region of the distribution. The critical Z score for a two-tailed test is written as Z_α. The critical Z score for a one-tailed test is written as $Z_{\alpha/2}$. The actual Z score is always compared to a critical Z score derived by observing a table. The actual Z score is calculated from the observed data using the test statistic

$$Z = \frac{\bar{x} - u}{\sigma/\sqrt{n}}$$

where n is the sample size.

Figure 3.1 depicts a standard normal distribution for a two-tailed hypothesis test with an alpha of 0.05. The rejection areas are the shaded areas at either end of the distribution; these two areas share the alpha. The null hypothesis would be rejected if the actual Z score calculated from the data is less than −1.960 or greater than 1.960. This could also be phrased as, "Reject the null hypothesis if the resulting absolute value of the actual Z score is greater than the absolute value of the critical Z score for the alpha." For example, a Z score of 1.210 would fall between −1.960 and 1.960, so we would fail to reject the null hypothesis. This could also be phrased as, "1.210 is less than $|1.960|$, so we fail to reject the null hypothesis."

Figure 3.2 depicts a standard normal distribution for a one-tailed lower tail hypothesis test with an alpha of 0.05. The rejection area is the shaded area at the lower tail of the distribution. The null hypothesis would be rejected if the resulting Z score is less than −1.645.

Figure 3.3 depicts a standard normal distribution for a one-tailed upper tail hypothesis test with an alpha of 0.05. The rejection area is the shaded area at the upper tail of the distribution. The null hypothesis would be rejected if the resulting actual Z score is greater than 1.645.

Certain assumptions must be met when performing a Z-test. The population must either be normal or have a sample size that is greater than a certain amount; the number 30 is often used. A minimum sample size of 30 is sufficient only if

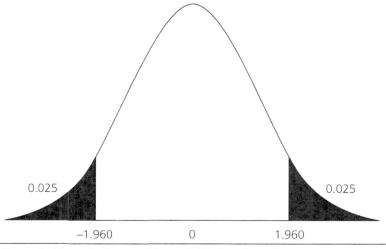

Figure 3.1 Two-tailed hypothesis test with $\alpha = 0.05$; therefore, $Z_\alpha = -1.960$ and 1.960.

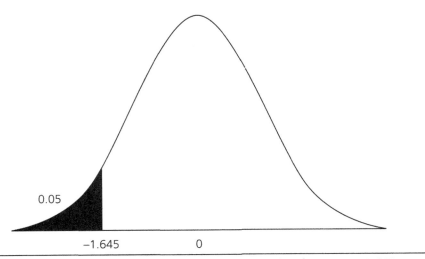

Figure 3.2   One-tailed lower tail hypothesis test with $\alpha = 0.05$; therefore, $Z_{\alpha/2} = -1.645$.

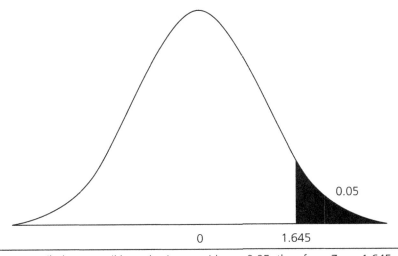

Figure 3.3   One-tailed upper tail hypothesis test with $\alpha = 0.05$; therefore, $Z_{\alpha/2} = 1.645$.

the data are nearly normal. If the data are highly non-normal, the sample size must be much larger than 30 if a Z-test is to be used. The population standard deviation must be known. The Z-test is used on a normal population. For a sample size of less than 30, the population must be normal; otherwise a Student's *t*-test (described below) should be used instead.

According to the central limit theorem, if you take multiple samples from a distribution that is not normally distributed and compute the mean for each sample, those means will approach a normal distribution. For example, if you have 500 values from a heavily skewed distribution and take 10 samples and compute their means, then repeat this until you have 9 sets of samples, the means of these samples will approach a normal distribution.

Note on Using Minitab for Hypothesis Tests

Minitab calculates a P value in place of using a Z score. A null hypothesis should be rejected if the resulting P value is less than 0.05 when using 95% confidence.

One-Sample Hypothesis Tests of Population Means

A one-sample hypothesis test is used to test a hypothesis regarding a sample average ($\bar{x}$) and a population average (μ). A one-sample hypothesis test could be used to determine if a sample average is representative of the population average.

Two-Tailed Z-Test

Assumptions: One sample with $n \geq 30$, any population shape, σ is known; or $n < 30$, population shape is normal, σ is known

Null hypothesis (H_0): $\mu = \mu_0$

Alternative hypothesis (H_a): $\mu \neq \mu_0$, with rejection criteria $Z < Z_{\alpha/2}$ or $Z > Z_{\alpha/2}$

Test statistic: $Z = \dfrac{\bar{x} - u}{\sigma/\sqrt{n}}$, with sample size $= n$

Example

A manufacturing company makes modifications to a production machine and believes the changes will have no result on the fatigue life of the parts produced by the machine. The standard deviation is known to be 18 hours. Prior to the modification, the parts lasted an average of 678 hours in a fatigue test. An engineer takes a sample of 32 parts and subjects them to a fatigue test; the sample average is found to be 674 hours.

Procedure

Step 1: Set the null hypothesis. The question is whether or not the process average is 678 hours, so a two-tailed hypothesis test is used.

$$H_0: \mu = 678$$

Step 2: Set the alternative hypothesis.

$$H_a: \mu \neq 678$$

Step 3: Set the alpha level and determine rejection criteria. We are using $\alpha = 0.05$, so reject if the test statistic is less than -1.96 or greater than 1.96.

Step 4: Determine the test statistic.

$$Z = \frac{674 - 678}{18/\sqrt{32}} = -1.26$$

Conclusion

In this case, the critical region is defined by values below –1.96 and above 1.96; –1.26 is not within the critical region, so we fail to reject the null hypothesis. There is insufficient evidence to support the alternative hypothesis that the average has changed since the modifications were implemented.

Minitab Instructions

Step 1: Go to "Stat > Basic Statistics > μ^Z 1-Sample Z…"

Step 2: Select "Summarized" data and enter the sample size, the mean, and the standard deviation.

Step 3: Check the box next to "Perform hypothesis test" and enter the hypothesized mean.

Step 4: Click on "Options…" and enter the desired confidence level and the alternative test. The default setting is 95.0 and "Mean ≠ hypothesized mean," which is what we need, so click OK to close the Options window.

Step 5: Click OK.

Minitab Output

One-Sample Z
Test of $\mu = 678$ vs $\neq 678$
The assumed standard deviation = 18

N	Mean	SE Mean	95% CI	Z	P
32	674.00	3.18	(667.76; 680.24)	–1.26	0.209

Interpretation of Minitab Output

We use the *P* value calculated by Minitab in place of the *Z* value calculated by hand. This Minitab-calculated *P* value of 0.209 is greater than an alpha of 0.05, so we fail to reject the null hypothesis.

One-Tailed Z-Test (Upper Tail)

Assumptions: One sample with $n \geq 30$, any population shape, σ is known; or $n < 30$, population shape is normal, σ is known

One-tailed null hypothesis (H_0): $\mu \leq \mu_0$

Alternative hypothesis (H_a): $\mu > \mu_0$, with rejection criteria $Z > Z_\alpha$

Test statistic: $Z = \dfrac{\bar{x} - u}{\sigma/\sqrt{n}}$, with sample size $= n$

Example

A Six Sigma Green Belt suspects customer complaints are costing on average $255 each. This is bad, but there is a possibility that the costs are much higher, so the Green Belt randomly selects 35 customer complaints for analysis and calculates a sample mean of $272. The standard deviation is calculated and found to be $14. The alpha is set at 0.05.

Procedure

Step 1: Set the null hypothesis. We want to know whether we can conclude that the average cost is higher than $255, so we use a one-tailed hypothesis with the upper tail.

$$H_0: \mu \leq 255.00$$

Step 2: Set the alternative hypothesis.

$$H_a: \mu > 255.00$$

Step 3: Set the alpha level and determine rejection criteria. We are using $\alpha = 0.05$, so reject if the test statistic is greater than 1.645.

Step 4: Determine the test statistic.

$$Z = \frac{272 - 255}{14/\sqrt{35}} = 7.18$$

Conclusion

Determine the rejection region for $\alpha = 0.05$. In this case the critical region is greater than 1.645; 7.18 is in the critical region, so we reject the null hypothesis. The evidence supports the alternative hypothesis that the average cost is greater than $255.

Minitab Instructions

Step 1: Go to "Stat > Basic Statistics > μZ 1-Sample Z…"

Step 2: Select "Summarized" data and enter the sample size, the mean, and the standard deviation.

Step 3: Check the box next to "Perform hypothesis test" and enter the hypothesized mean.

Step 4: Click on "Options…" and enter the desired confidence level and the alternative test. The default setting is 95.0 and "Mean > hypothesized mean," which is what we need, so click OK to close the Options window.

Step 5: Click OK.

Minitab Output

One-Sample Z
Test of $\mu = 255$ vs > 255
The assumed standard deviation = 14

N	Mean	SE Mean	95% Lower Bound	Z	P
35	272.00	2.37	268.11	7.18	0.000

Interpretation of Minitab Output

We use the P value calculated by Minitab in place of the Z value calculated by hand. This Minitab-calculated P value of 0.000 is less than an alpha of 0.05, so we reject the null hypothesis in favor of the alternative hypothesis.

One-Tailed Z-Test (Lower Tail)

Assumptions: One sample with $n \geq 30$, any population shape, σ is known; or $n < 30$, population shape is normal, σ is known

One-tailed null hypothesis (H_0): $\mu \geq \mu_0$

Alternative hypothesis (H_a): $\mu < \mu_0$, with rejection criteria $Z < Z_\alpha$

Test statistic: $Z = \dfrac{\bar{x} - u}{\sigma/\sqrt{n}}$, with sample size $= n$

Example

Suppose the random sampling of 35 customer complaints in the previous example resulted in a sample mean of only \$247, which means the average customer complaint may cost less than \$255. The standard deviation is still \$14 and the alpha is set at 0.05. Here we still use a one-tailed hypothesis test, but this time it is for the lower tail.

Procedure

Step 1: Set the null hypothesis. We want to know whether we can conclude that the average cost is lower than \$255, so we use a one-tailed hypothesis with the lower tail.

$$H_0: \mu \geq 255.00$$

Step 2: Set the alternative hypothesis.

$$H_a: \mu < 255.00$$

Step 3: Set the alpha level and determine rejection criteria. We are using $\alpha = 0.05$, so reject if the test statistic is less than -1.645.

Step 4: Determine the test statistic.

$$Z = \frac{247 - 255}{14/\sqrt{35}} = -3.38$$

Conclusion

In this case the critical region is less than -1.645; -3.38 is in the critical region, so we reject the null hypothesis. The evidence supports the alternative hypothesis that the average cost is less than \$255.

Minitab Instructions

Step 1: Go to "Stat > Basic Statistics > μZ 1-Sample Z…"

Step 2: Select "Summarized" data and enter the sample size, the mean, and the standard deviation.

Step 3: Check the box next to "Perform hypothesis test" and enter the hypothesized mean.

Step 4: Click on "Options…" and enter the desired confidence level and the alternative test. The default setting is 95.0 and "Mean < hypothesized mean," which is what we need, so click OK to close the Options window.

Step 5: Click OK.

Minitab Output

```
One-Sample Z
Test of μ = 255 vs < 255
The assumed standard deviation = 14
```

N	Mean	SE Mean	95% Upper Bound	Z	P
35	247.00	2.37	250.89	−3.38	0.000

Interpretation of Minitab Output

The P value calculated by Minitab is less than the P value of $\alpha = 0.05$, so we fail to reject the null hypothesis.

t-Tests

The Z-test is used when the sample size is 30 or larger or the population has a normal distribution; the Student's t distribution, or t-test, is used when the sample size is less than 30 and the distribution is not normal. Three typical t-tests are the one-sample hypothesis test about the population mean, the two-sample population test of independent means, and the paired t-test. The one-sample t-test is used to compare a sample from a population against a hypothesized mean. The two-sample t test of independent means is used to compare two samples to each other. It is analogous to an ANOVA but uses only two levels of the categorical variable. The t-test for paired samples compares the means of defined pairs, such as when comparing before and after results.

The Z-test uses the known standard deviation, denoted by σ. The Student's t distribution uses the sample standard deviation, denoted by s; it is larger than the population parameter owing to some uncertainty in its estimation. The resulting bell-shaped curve for s is wider than the bell-shaped curve for σ because of this increased uncertainty (Johnson and Bhattacharyya 2010). Student's t distribution is named after Guinness brewery employee William Sealy Gosset, who published under the pseudonym "Student" so that competitors would not know his employer was using statistical methods (Zilak 2008).

The t distribution has "heavier" tails than the standard normal distribution (Lawson and Erjavec 2001); however, the t distribution approaches the standard normal distribution as the number of degrees of freedom increases. The number of degrees of freedom for the t distribution is determined by subtracting 1 from the sample size, so degrees of freedom = $n - 1$.

Figure 3.4 depicts the standard normal distribution compared to a t distribution with 5 degrees of freedom. Figure 3.5 shows a standard normal distribution and a t distribution with 15 degrees of freedom. The t distribution starts to match the standard normal distribution as sample size, and therefore degrees of freedom, increases.

The t statistic is determined by looking at a t distribution table such as the one in Appendix C. The degrees of freedom are located in the left column of the table, and the t statistic is found at the point where the degrees of freedom meet the required alpha level. For example, suppose the sample size is 4; this means the degrees of freedom are equal to 4 minus 1. Looking at the t table in Figure 3.6, we see that 3 degrees of freedom with an alpha of 0.05 for a two-tailed test is 3.182.

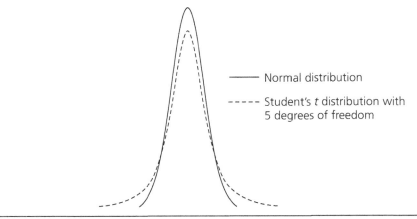

Figure 3.4   Standard normal distribution and *t* distribution with 5 degrees of freedom.

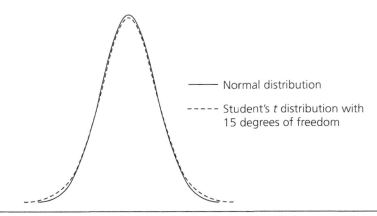

Figure 3.5   Standard normal distribution and *t* distribution with 15 degrees of freedom.

		Two-tailed probabilities		
		0.10	0.05	0.01
	1	6.314	12.71	63.66
	2	2.920	4.303	9.925
Degrees of freedom	**3**	2.353	3.182	5.841
	4	2.132	2.776	4.604
	5	2.015	2.571	4.032
	6	1.943	2.447	3.707

Figure 3.6   Example of a *t* distribution table.

One-Sample Hypothesis Tests about the Population Mean (t-Test)

Two-Tailed t-Test

Assumptions: One sample with $n < 30$, population is normally distributed, standard deviation is unknown

Null hypothesis (H_0): $\mu = \mu_0$

Alternative hypothesis (H_a): $\mu \neq \mu_0$, with rejection criteria $t < t_{\alpha/2,\,n-1}$ or $t > t_{\alpha/2,\,n-1}$

Test statistic: $t = \dfrac{\bar{x} - \mu}{s/\sqrt{n}}$; degrees of freedom $= n - 1$

Example

The diameter of machined components should have a process mean of 28 mm. A Six Sigma Black Belt selects a sample of 20 parts and measures them. The sample mean is 27.8 mm. The standard deviation for the sample is 0.8. The Six Sigma Black Belt needs to determine if the population mean is equal to 28 mm.

Procedure

Step 1: Set the null hypothesis. The question is whether or not the true population mean is 28 mm, so a two-tailed hypothesis test is used.

$$H_0: \mu = 28$$

Step 2: Set the alternative hypothesis.

$$H_a: \mu \neq 28$$

Step 3: Set the alpha level and determine rejection criteria. We are using $\alpha = 0.05$ with $n - 1 = 19$ degrees of freedom, so we look at the t distribution table and determine t is 2.093 for a two-tailed test with 19 degrees of freedom and an alpha of 0.05.

Step 4: Determine the test statistic.

$$t = \frac{27.8 - 28}{0.8/\sqrt{20}} = -1.12$$

Conclusion

In this case, the critical region is left of −2.093 and right of 2.093. The actual t score of −1.12 is not within the critical region, so we fail to reject the null hypothesis. There is insufficient evidence to support the alternative hypothesis that the true mean is not 28 mm.

Minitab Instructions

Step 1: Go to "Stat > Basic Statistics > μ 1-Sample t…"

Step 2: Select "Summarized" data and enter the sample size, the mean, and the standard deviation.

Step 3: Check the box next to "Perform hypothesis test" and enter the hypothesized mean.

Step 4: Click on "Options…" and enter the desired confidence level and the alternative test. The default setting is 95.0 and "Mean ≠ hypothesized mean," which is what we need, so click OK to close the Options window.

Step 5: Click OK.

Minitab Output

One-Sample T						
Test of μ = 28 vs ≠ 28						
N	Mean	StDev	SE Mean	95% CI	T	P
20	27.800	0.800	0.179	(27.426; 28.174)	−1.12	0.277

Interpretation of Minitab Output

We use the *P* value calculated by Minitab in place of the *t* value calculated using the Student's *t* distribution. This Minitab-calculated *P* value of 0.277 is greater than an alpha of 0.05, so we fail to reject the null hypothesis.

One-Tailed t-*Test (Upper Tail)*

Assumptions: One sample with $n < 30$, population is normally distributed, standard deviation is unknown

Null hypothesis (H_0): $\mu \leq \mu_0$

Alternative hypothesis (H_a): $\mu > \mu_0$, with rejection criteria $t > t_{\alpha, n-1}$

Test statistic: $t = \dfrac{\bar{x} - \mu}{s/\sqrt{n}}$; degrees of freedom = $n - 1$

Example

Suppose the Six Sigma Black Belt in the previous example used a sample of 12 parts; the sample mean is 27.4 mm and the sample standard deviation is 1.1. The process mean is thought to be 28 mm, so the Six Sigma Black Belt decides to construct a hypothesis test to determine if the population mean is actually greater than 28.

Procedure

Step 1: Set the null hypothesis. The question is whether or not the population mean is greater than 28 mm, so a one-tailed upper tail test is used.

$$H_0: \mu \leq 28$$

Step 2: Set the alternative hypothesis.

$$H_a: \mu > 28$$

Step 3: Set the alpha level and determine rejection criteria. We are using $\alpha = 0.05$ with $n - 1 = 11$ degrees of freedom, so we look at the *t* distribution table and determine *t* critical for a one-tailed upper tail test with 11 degrees of freedom and an alpha of 0.05. The critical region is above 1.796.

Step 4: Determine the test statistic.

$$t = \frac{27.4 - 28}{1.1/\sqrt{12}} = -1.89$$

Conclusion

In this case, the critical region lies above 1.796; because −1.89 is not within the critical region, we fail to reject the null hypothesis. The evidence does not support the alternative hypothesis that the true mean is greater than 28 mm.

Minitab Instructions

Step 1: Go to "Stat > Basic Statistics > μ 1-Sample t…"

Step 2: Select "Summarized" data and enter the sample size, the mean, and the standard deviation.

Step 3: Check the box next to "Perform hypothesis test" and enter the hypothesized mean.

Step 4: Click on "Options…" and enter the desired confidence level and the alternative test. The default setting is 95.0 and "Mean > hypothesized mean," which is what we need, so click OK to close the Options window.

Step 5: Click OK.

Minitab Output

One-Sample T
Test of μ = 28 vs > 28

N	Mean	StDev	SE Mean	95% Lower Bound	T	P
12	27.400	1.100	0.318	26.830	−1.89	0.957

Interpretation of Minitab Output

We use the P value calculated by Minitab in place of the t value calculated using the Student's t distribution. This Minitab-calculated P value of 0.957 is greater than the alpha of 0.05, so we fail to reject the null hypothesis.

One-Tailed t-*Test (Lower Tail)*

> Assumptions: One sample with $n < 30$, population is normally distributed, standard deviation is unknown
>
> Null hypothesis (H_0): $\mu \geq \mu_0$
>
> Alternative hypothesis (H_a): $\mu < \mu_0$, with rejection criteria $t < t_{\alpha, n-1}$
>
> Test statistic: $t = \dfrac{\bar{x} - \mu}{s/\sqrt{n}}$; degrees of freedom = $n - 1$

Example

Suppose the sample of 20 parts in the earlier example was 28.6 mm with a standard deviation of 1.1. The Six Sigma Black Belt wants to determine if there is evidence that the true population mean is less than a hypothesized process mean of 28 mm.

Procedure

Step 1: Set the null hypothesis. The question is whether or not the population mean is less than 28 mm, so a one-tailed lower tail test is used.

$$H_0: \mu \geq 28$$

Step 2: Set the alternative hypothesis.

$$H_a: \mu < 28$$

Step 3: Set the alpha level and determine rejection criteria. We are using $\alpha = 0.05$ with $n - 1 = 19$ degrees of freedom, so we look at the t distribution table and determine that the rejection region for a one-tailed lower tail test with 19 degrees of freedom and an alpha of 0.05 is -1.729.

Step 4: Determine the test statistic.

$$t = \frac{28.6 - 28}{1.1/\sqrt{20}} = 2.44$$

Conclusion

In this case, the critical region is less than -1.729; because 2.44 is not within the critical region, we fail to reject the null hypothesis. There is insufficient evidence to support the alternative hypothesis that the true mean is less than 28 mm.

Minitab Instructions

Step 1: Go to "Stat > Basic Statistics > μ 1-Sample t…"

Step 2: Select "Summarized" data and enter the sample size, the mean, and the standard deviation.

Step 3: Check the box next to "Perform hypothesis test" and enter the hypothesized mean.

Step 4: Click on "Options…" and enter the desired confidence level and the alternative test. The default setting is 95.0 and "Mean < hypothesized mean," which is what we need, so click OK to close the Options window.

Step 5: Click OK.

Minitab Output

One-Sample T
Test of $\mu = 28$ vs < 28

N	Mean	StDev	SE Mean	95% Upper Bound	T	P
20	28.600	1.100	0.246	29.025	2.44	0.988

Interpretation of Minitab Output

We use the P value calculated by Minitab in place of the t value calculated using the Student's t distribution. This Minitab-calculated P value of 0.988 is greater than the alpha of 0.05, so we fail to reject the null hypothesis in favor of the alternative hypothesis.

Two-Sample Hypothesis Tests of Means with Unequal Variances (t-Test)

A two-sample Student's *t*-test is used to compare samples from two separate populations when the sample standard deviations are unknown and unequal.

Two-Tailed t-*Test*

Assumptions: n_1. $n_2 < 30$, independent samples, population is normally distributed, standard deviations are unknown and unequal

Null hypothesis (H_0): $\mu_1 - \mu_2 = 0$

Alternative hypothesis (H_a): $\mu_1 - \mu_2 \neq 0$, with rejection criteria $t < t_{\alpha/2,\, n-1}$ or $t > t_{\alpha/2,\, n-1}$

Test statistic: $t = \dfrac{(\bar{x}_1 - \bar{x}_2) - (\mu_1 - \mu_2)}{\sqrt{\dfrac{s_1^2}{n_1} + \dfrac{s_2^2}{n_2}}}$; degrees of freedom $= \dfrac{\left(\dfrac{s_1^2}{n_1} + \dfrac{s_2^2}{n_2}\right)^2}{\dfrac{\left(\dfrac{s_1^2}{n_1}\right)^2}{n_1 - 1} + \dfrac{\left(\dfrac{s_2^2}{n_2}\right)^2}{n_2 - 1}}$

Example

A steel tube is cut to the required length on either of two different recut machines. The production supervisor wants to know if the parts produced on both machines have the same mean, so she asks a Six Sigma Green Belt to investigate. The quality technician collects 20 samples from machine one (n_1) and 18 samples from machine two (n_2). The quality technician determines that the mean of machine one ($\bar{x}_1$) is 854 mm and the mean of machine two ($\bar{x}_2$) is 867 mm. The standard deviation of machine one (s_1) is 10.3 and the standard deviation of machine two (s_2) is 6.43.

Procedure

Step 1: Set the null hypothesis. The question is whether or not the difference in the means of the two machines is zero, so a two-tailed hypothesis test is used.

$$H_0\colon \mu_1 - \mu_2 = 0$$

Step 2: Set the alternative hypothesis.

$$H_a\colon \mu_1 - \mu_2 \neq 0$$

Step 3: Set the alpha level and determine rejection criteria. We are using $\alpha = 0.05$ and must calculate the degrees of freedom.

$$\frac{\left(\dfrac{10.3^2}{20} + \dfrac{6.4^2}{18}\right)^2}{\dfrac{\left(\dfrac{10.3^2}{20}\right)^2}{20 - 1} + \dfrac{\left(\dfrac{6.4^2}{18}\right)^2}{18 - 1}} = 32.180$$

Thus, we round down and use 32 degrees of freedom.

We then look at the *t* distribution table and determine the *t* value for a two-tailed test with 32 degrees of freedom and an alpha of 0.05. The table does not give the exact value for 32 degrees of freedom, so we round down to 30 and use 2.042.

Step 4: Determine the test statistic. We believe there is no difference between the means, so we set $\mu_1 - \mu_2$ equal to 0.

$$t = \frac{(854 - 867) - (0)}{\sqrt{\frac{10.3^2}{20} + \frac{6.4^2}{18}}} = -4.72$$

Conclusion

The rejection region is lower than −2.042 or higher than 2.042, and −4.72 is less than −2.045, so we reject the null hypothesis. The evidence supports the alternative hypothesis that the population means are different.

Minitab Instructions

Step 1: Go to "Stat > Basic Statistics > μ 2-Sample t…"

Step 2: Select "Summarized" data and enter the sample size, the mean, and the standard deviation.

Step 3: Click on "Options…" and enter the desired confidence level. The default setting is 95.0. The "Hypothesized difference" is zero, so we leave that as 0.0. Select "Difference ≠ hypothesized difference," then click OK to close the Options window.

Step 4: Click OK.

Minitab Output

Two-Sample T-Test and CI

Sample	N	Mean	StDev	SE Mean
1	20	854.0	10.3	2.3
2	18	867.00	6.43	1.5

Difference = μ (1) − μ (2)

Estimate for difference: −13.00

95% CI for difference: (−18.62; −7.38)

T-Test of difference = 0 (vs ≠): T-Value = −4.72
P-Value = 0.000 DF = 32

Interpretation of Minitab Output

We use the *P* value calculated by Minitab in place of the *t* value calculated using the Student's *t* distribution. This Minitab-calculated *P* value of 0.000 is less than the alpha of 0.05, so we reject the null hypothesis in favor of the alternative hypothesis that the population means are different.

One-Tailed t-*Test (Upper Tail)*

Assumptions: n_1. $n_2 < 30$, independent samples, population is normally distributed, standard deviations are unknown and unequal

Null hypothesis (H_0): $\mu_1 - \mu_2 \leq \delta$, where δ is the hypothesized difference between μ_1 and μ_2

Alternative hypothesis (H_a): $\mu_1 - \mu_2 > \delta$, with rejection criteria $t > t_{\alpha,\,n-1}$

Test statistic: $t = \dfrac{(\bar{x}_1 - \bar{x}_2) - (\mu_1 - \mu_2)}{\sqrt{\dfrac{s_1^2}{n_1} + \dfrac{s_2^2}{n_2}}}$; degrees of freedom $= \dfrac{\left(\dfrac{s_1^2}{n_1} + \dfrac{s_2^2}{n_2}\right)^2}{\dfrac{\left(\dfrac{s_1^2}{n_1}\right)^2}{n_1 - 1} + \dfrac{\left(\dfrac{s_2^2}{n_2}\right)^2}{n_2 - 1}}$

Example

Suppose the Six Sigma Green Belt from the last example had actually found a mean of 862 mm with a standard deviation of 8.3 for 24 samples from machine one, and a mean of 858 mm with a standard deviation of 7.3 for 21 samples from machine two. The Six Sigma Green Belt wants to know if the difference between the two machines is actually more than 5.

Procedure

Step 1: Set the null hypothesis. A one-tailed upper tail test is used to determine whether the difference in means is greater than or equal to 5.

$$H_0: \mu_1 - \mu_2 \leq 5$$

Step 2: Set the alternative hypothesis.

$$H_a: \mu_1 - \mu_2 > 5$$

Step 3: Set the alpha level and determine rejection criteria. We are using $\alpha = 0.05$ and must calculate the degrees of freedom.

$$\frac{\left(\dfrac{8.3^2}{24} + \dfrac{7.3^2}{21}\right)^2}{\dfrac{\left(\dfrac{8.3^2}{24}\right)^2}{24 - 1} + \dfrac{\left(\dfrac{7.3^2}{21}\right)^2}{21 - 1}} = 42 \text{ degrees of freedom}$$

We then look at the t distribution table and determine the t value for a one-tailed test with 42 degrees of freedom and an alpha of 0.05. The table does not give the exact value for 42 degrees of freedom, so we round down to 40 and use 2.021.

Step 4: Determine the test statistic. We believe there is a difference of 5 between the means, so we set $\mu_1 - \mu_2$ equal to 5.

$$t = \frac{(862 - 858) - (5)}{\sqrt{\dfrac{8.3^2}{24} + \dfrac{7.3^2}{21}}} = -0.43$$

Conclusion

The rejection region is higher than 2.021, and –0.43 is less than 2.021, so we fail to reject the null hypothesis. The evidence does not support the alternative hypothesis that the difference in means is greater than 5.

Minitab Instructions

Step 1: Go to "Stat > Basic Statistics > μ 2-Sample t…"

Step 2: Select "Summarized" data and enter the sample size, the mean, and the standard deviation.

Step 3: Click on "Options…" and enter the desired confidence level. The default setting is 95.0. Under "Hypothesized difference," enter a 5. Select "Difference > hypothesized difference," then click OK to close the Options window.

Step 4: Click OK.

Minitab Output

Two-Sample T-Test and CI

Sample	N	Mean	StDev	SE Mean
1	24	862.00	8.30	1.7
2	21	858.00	7.30	1.6

Difference = μ (1) − μ (2)

Estimate for difference: 4.00

95% lower bound for difference: 0.09

T-Test of difference = 5 (vs >): T-Value = −0.43
P-Value = 0.665 DF = 42

Interpretation of Minitab Output

We use the *P* value calculated by Minitab in place of the *t* value calculated using the Student's *t* distribution. This Minitab-calculated *P* value of 0.665 is greater than the alpha of 0.05, so we fail to reject the null hypothesis.

One-Tailed t-*Test (Lower Tail)*

Assumptions: n_1. $n_2 < 30$, independent samples, population is normally distributed, standard deviations are unknown and unequal

Null hypothesis (H_0): $\mu_1 - \mu_2 \geq \delta$, where δ is the hypothesized difference between μ_1 and μ_2

Alternative hypothesis (H_a): $\mu_1 - \mu_2 < \delta$, with rejection criteria $t < t_{\alpha, n-1}$

Test statistic: $t = \dfrac{(\bar{x}_1 - \bar{x}_2) - (\mu_1 - \mu_2)}{\sqrt{\dfrac{s_1^2}{n_1} + \dfrac{s_2^2}{n_2}}}$; degrees of freedom $= \dfrac{\left(\dfrac{s_1^2}{n_1} + \dfrac{s_2^2}{n_2}\right)^2}{\dfrac{\left(\dfrac{s_1^2}{n_1}\right)^2}{n_1 - 1} + \dfrac{\left(\dfrac{s_2^2}{n_2}\right)^2}{n_2 - 1}}$

Example

Using the details from an earlier example, n_1 is 20 and n_2 is 18. The sample standard deviations are 10.3 for s_1 and 6.43 for s_2. The sample mean for $\bar{x}_1$ is 854 mm and $\bar{x}_2$ is 867 mm.

There was insufficient evidence to conclude that the population means were equal, so now the Six Sigma Green Belt wants to determine if the mean of $\bar{x}_1$ is actually less than 10 lower than $\bar{x}_2$.

Procedure

Step 1: Set the null hypothesis. A one-tailed lower tail test is used to determine whether the difference in means is less than or equal to 10. The mean of μ_1 is believed to be lower than the mean of μ_2, so the difference must be a negative number.

$$H_0: \mu_1 - \mu_2 \geq -10$$

Step 2: Set the alternative hypothesis.

$$H_a: \mu_1 - \mu_2 < -10$$

Step 3: Set the alpha level and determine rejection criteria. We are using $\alpha = 0.05$ and must calculate the degrees of freedom.

$$\frac{\left(\frac{10.3^2}{20} + \frac{6.4^2}{18}\right)^2}{\frac{\left(\frac{10.3^2}{20}\right)^2}{20 - 1} + \frac{\left(\frac{6.4^2}{18}\right)^2}{18 - 1}} = 32.180$$

Thus, we round down and use 32 degrees of freedom. We then look at the t distribution table and determine the t value for a one-tailed test with 32 degrees of freedom and an alpha of 0.05. The table does not give the exact value for 32 degrees of freedom, so we round down to 30 and use –2.042.

Step 4: Determine the test statistic. We believe there is a difference of 10 between the means, so we set $\mu_1 - \mu_2$ equal to –10.

$$t = \frac{(854 - 867) - (-10)}{\sqrt{\frac{10.3^2}{20} + \frac{6.4^2}{18}}} = -8.35$$

Conclusion

The rejection region is lower than –2.042, and –8.35 is less than –2.042, so we reject the null hypothesis. The evidence supports the alternative hypothesis that the difference in the means of the two populations is less than 10.

Minitab Instructions

Step 1: Go to "Stat > Basic Statistics > μ 2-Sample t…"

Step 2: Select "Summarized" data and enter the sample size, the mean, and the standard deviation.

Step 3: Click on "Options…" and enter the desired confidence level. The default setting is 95.0. Under "Hypothesized difference," enter a 10. Select "Difference < hypothesized difference," then click OK to close the Options window.

Step 4: Click OK.

Minitab Output

Two-Sample T-Test and CI

Sample	N	Mean	StDev	SE Mean
1	20	854.0	10.3	2.3
2	18	867.00	6.43	1.5

Difference = $\mu(1) - \mu(2)$

Estimate for difference: -13.00

95% upper bound for difference: -8.33

T-Test of difference = 10 (vs <): T-Value = -8.34
P-Value = 0.000 DF = 32

Interpretation of Minitab Output

We use the *P* value calculated by Minitab in place of the *t* value calculated using the Student's *t* distribution. This Minitab-calculated *P* value of 0.000 is less than the alpha of 0.05, so we reject the null hypothesis in favor of the alternative hypothesis.

Two-Sample Hypothesis Tests of Means with Equal Variances (t-Test)

A different type of two-sample Student's *t*-test is used to compare samples from two independent samples when the sample standard deviations are unknown but equal.

Two-Tailed t-*Test*

Assumptions: n_1. $n_2 < 30$, independent samples, population is normally distributed, standard deviations are unknown and equal

Null hypothesis (H_0): $\mu_1 - \mu_2 = 0$

Alternative hypothesis (H_a): $\mu_1 - \mu_2 \neq 0$, with rejection criteria $t < t_{\alpha/2,\, n1+n2-2}$ or $t > t_{\alpha/2,\, n1+n2-2}$

Test statistic: $t = \dfrac{(\bar{x}_1 - \bar{x}_2) - (\mu_1 - \mu_2)}{\sqrt{\dfrac{(n_1 - 1)s_1^2 + (n_2 - 1)s_2^2}{n_1 + n_2 - 2}}\sqrt{\dfrac{1}{n_1} + \dfrac{1}{n_2}}}$; degrees of freedom = $n_1 + n_2 - 2$

Example

A sales manager samples 24 customers from sales region one and 26 customers from sales region two and determines that in one month, the customers in sales region one purchased material with a mean value of $1400 with a standard deviation of 118, while purchases in sales region two had a mean of $1350 with a standard deviation of 116. The sales manager needs to determine whether the means of the two regions are equal.

Procedure

Step 1: Set the null hypothesis. The question is whether or not the difference in the means of the two sales regions is zero, so a two-tailed hypothesis test is used.

$$H_0: \mu_1 - \mu_2 = 0$$

Step 2: Set the alternative hypothesis.

$$H_a: \mu_1 - \mu_2 \neq 0$$

Step 3: Set the alpha level and determine rejection criteria. We are using $\alpha = 0.05$ and must calculate the degrees of freedom, which is $n_1 + n_2 - 2$, so $24 + 26 - 2 = 48$. We then look at the t distribution table and determine the t value for a two-tailed test with 48 degrees of freedom and an alpha of 0.05. The value is 2.011.

Step 4: Determine the test statistic. We believe there is no difference between the means, so we set $\mu_1 - \mu_2$ equal to 0.

$$t = \frac{(1400 - 1350) - (10)}{\sqrt{\dfrac{(24 - 1)118^2 + (26 - 1)116^2}{24 + 26 - 2}} \sqrt{\dfrac{1}{24} + \dfrac{1}{26}}} = 1.51$$

Conclusion

The rejection region is lower than –2.011 or higher than 2.011, and 1.51 is within the acceptance region, so we fail to reject the null hypothesis. The evidence supports the alternative hypothesis that there is a difference between the two sales regions.

Minitab Instructions

Step 1: Go to "Stat > Basic Statistics > μ 2-Sample t..."

Step 2: Select "Summarized" data and enter the sample size, the mean, and the standard deviation.

Step 3: Click on "Options..." and enter the desired confidence level. The default setting is 95.0. The "Hypothesized difference" is zero, so we leave that as 0.0. Select "Difference ≠ hypothesized difference," then click OK to close the Options window.

Step 4: Check the box marked "Assume equal variances."

Step 5: Click OK.

Minitab Output

Two-Sample T-Test and CI

Sample	N	Mean	StDev	SE Mean
1	24	1400	118	24
2	26	1350	116	23

Difference = $\mu(1) - \mu(2)$

Estimate for difference: 50.0

95% CI for difference: (−16.6; 116.6)

T-Test of difference = 0 (vs ≠): T-Value = 1.51
P-Value = 0.138 DF = 48

Both use Pooled StDev = 116.9626

Interpretation of Minitab Output

We use the *P* value calculated by Minitab in place of the *t* value calculated using the Student's *t* distribution. This Minitab-calculated *P* value of 0.138 is greater than the alpha of 0.05, so we fail to reject the null hypothesis.

One-Tailed t-*Test (Upper Tail)*

Assumptions: n_1. $n_2 < 30$, independent samples, population is normally distributed, standard deviations are unknown and equal

Null hypothesis (H_0): $\mu_1 - \mu_2 \leq \delta$, where δ is the hypothesized difference between μ_1 and μ_2

Alternative hypothesis (H_a): $\mu_1 - \mu_2 > \delta$, with rejection criteria $t > t_{\alpha, n1+n2-2}$

Test statistic: $t = \dfrac{(\bar{x}_1 - \bar{x}_2) - (\mu_1 - \mu_2)}{\sqrt{\dfrac{(n_1-1)s_1^2 + (n_2-1)s_2^2}{n_1 + n_2 - 2}}\sqrt{\dfrac{1}{n_1} + \dfrac{1}{n_2}}}$; degrees of freedom = $n_1 + n_2 - 2$

Example

Suppose the sales manager in the previous example wants to determine if the mean of the sales in region one is more than $100 more than the mean of region two. The sample size for region one is 24 and the mean is $1400 with a standard deviation of 118. The mean of region two is $1350 with a sample size of 26 and a standard deviation of 116.

Procedure

Step 1: Set the null hypothesis. A one-tailed upper tail test is used to determine whether the difference in means is greater than or equal to 100.

$$H_0: \mu_1 - \mu_2 \leq 100$$

Step 2: Set the alternative hypothesis.

$$H_a: \mu_1 - \mu_2 > 100$$

Step 3: Set the alpha level and determine rejection criteria. We are using $\alpha = 0.05$ and must calculate the degrees of freedom, which is $n_1 + n_2 - 2$, so $24 + 26 - 2 = 48$. We then look at the t distribution table and determine the t value for a one-tailed test with 48 degrees of freedom and an alpha of 0.05. The table does not give the exact value for 48 degrees of freedom, so we round down to 40 and use 1.684.

Step 4: Determine the test statistic. We believe there is a difference of 100 between the means, so we set $\mu_1 - \mu_2$ equal to 100.

$$t = \frac{(1400 - 1350) - (10)}{\sqrt{\frac{(24 - 1)118^2 + (26 - 1)116^2}{24 + 26 - 2}} \sqrt{\frac{1}{24} + \frac{1}{26}}} = 1.51$$

Conclusion

The rejection region is higher than 1.684, and –1.51 is less than 1.684, so we fail to reject the null hypothesis. The evidence does not support the alternative hypothesis that the difference in means is greater than 100.

Minitab Instructions

Step 1: Go to "Stat > Basic Statistics > μ 2-Sample t..."

Step 2: Select "Summarized" data and enter the sample size, the mean, and the standard deviation.

Step 3: Click on "Options..." and enter the desired confidence level. The default setting is 95.0. Under "Hypothesized difference," enter 100. Select "Difference > hypothesized difference," then click OK to close the Options window.

Step 4: Check the box marked "Assume equal variances."

Step 5: Click OK.

Minitab Output

Two-Sample T-Test and CI

Sample	N	Mean	StDev	SE Mean
1	24	1400	118	24
2	26	1350	116	23

Difference = μ (1) – μ (2)

Estimate for difference: 50.0

95% lower bound for difference: –5.5

T-Test of difference = 100 (vs >): T-Value = –1.51
P-Value = 0.931 DF = 48

Both use Pooled StDev = 116.9626

Interpretation of Minitab Output

We use the *P* value calculated by Minitab in place of the *t* value calculated using the Student's *t* distribution. This Minitab-calculated *P* value of 0.931 is greater than the alpha of 0.05, so we fail to reject the null hypothesis.

One-Tailed t-Test (Lower Tail)

Assumptions: n_1. $n_2 < 30$, independent samples, population is normally distributed, standard deviations are unknown and equal

Null hypothesis (H_c): $\mu_1 - \mu_2 \geq \delta$, where δ is the hypothesized difference between μ_1 and μ_2

Alternative hypothesis (H_a): $\mu_1 - \mu_2 < \delta$, with rejection criteria $t < t_{\alpha,\, n1+n2-2}$

Test statistic: $t = \dfrac{(\bar{x}_1 - \bar{x}_2) - (\mu_1 - \mu_2)}{\sqrt{\dfrac{(n_1 - 1)s_1^2 + (n_2 - 1)s_2^2}{n_1 + n_2 - 2}}\sqrt{\dfrac{1}{n_1} + \dfrac{1}{n_2}}}$; degrees of freedom $= n_1 + n_2 - 2$

Example

Suppose the sales manager in the previous example compares sales region one (with a sample size of 24, a mean of $1400, and a standard deviation of 118) to sales region three instead of sales region two. The sample size from sales region three is 19, the mean is found to be $1300, and the standard deviation is 119. The sales manager wants to know if the difference between the two means is less than $200.

Procedure

Step 1: Set the null hypothesis. A one-tailed lower tail test is used to determine if the difference in means is less than or equal to 200. The mean of μ_1 is believed to be lower than the mean of μ_2, so the difference must be a negative number.

$$H_0: \mu_1 - \mu_2 \geq 200$$

Step 2: Set the alternative hypothesis.

$$H_a: \mu_1 - \mu_2 < 200$$

Step 3: Set the alpha level and determine rejection criteria. We are using $\alpha = 0.05$ and must calculate the degrees of freedom, which is $n_1 + n_2 - 2$, so $24 + 19 - 2 = 41$. We then look at the *t* distribution table and determine the *t* value for a one-tailed test with 41 degrees of freedom and an alpha of 0.05. The table does not give the exact value for 41 degrees of freedom, so we round down to 40 and use −1.684.

Step 4: Determine the test statistic. We believe there is a difference of 200 between the means, so we set $\mu_1 - \mu_2$ equal to 200.

$$t = \frac{(1400 - 1300) - (200)}{\sqrt{\dfrac{(24 - 1)118^2 + (19 - 1)119^2}{24 + 19 - 2}}\sqrt{\dfrac{1}{24} + \dfrac{1}{19}}} = -2.75$$

Conclusion

The rejection region is lower than −1.684, and −2.75 is less than −1.684, so we reject the null hypothesis. The evidence supports the alternative hypothesis that the difference in means is less than 200.

Minitab Instructions

Step 1: Go to "Stat > Basic Statistics > μ 2-Sample t…"

Step 2: Select "Summarized" data and enter the sample size, the mean, and the standard deviation.

Step 3: Click on "Options…" and enter the desired confidence level. The default setting is 95.0. Under "Hypothesized difference," enter a 10. Select "Difference < hypothesized difference," then click OK to close the Options window.

Step 4: Check the box marked "Assume equal variances."

Step 5: Click OK.

Minitab Output

Two-Sample T-Test and CI

Sample	N	Mean	StDev	SE Mean
1	24	1400	118	24
2	19	1300	119	27

Difference = μ (1) − μ (2)

Estimate for difference: 100.0

95% upper bound for difference: 161.2

T-Test of difference = 200 (vs <): T-Value = −2.75
P-Value = 0.004 DF = 41

Both use Pooled StDev = 118.4401

Interpretation of Minitab Output

We use the P value calculated by Minitab in place of the t value using the Student's t distribution. This Minitab-calculated P value of 0.004 is greater than the alpha of 0.05, so we reject the null hypothesis.

Paired **t-Tests**

Paired t-tests are used to compare dependent samples that are paired together. The samples must be collected under comparable conditions (Montgomery, Runger, and Hubele 2001), such as before and after the results of an individual or individuals have been paired together because they share similar characteristics. The hypothesis tests for paired t-test use elements of one data set paired to an element of the other data set and a normal distribution. These tests are useful for assessing improvements to a process or machine when before-and-after data are available.

Two-Tailed Paired t-*Test*

Assumptions: Sample size $n \le 30$, sample sizes are equal, population is normally distributed, standard deviations are approximately equal. Each population is tested twice, with observations paired together; the observations within the populations must be independent of each other. In a paired t-test, μ_d is the difference in means and μ_0 is the hypothesized difference between the means.

Null hypothesis (H_0): $\mu_d = \mu_0$

Alternative hypothesis (H_a): $\mu_d \ne \mu_0$, with rejection criteria $t < -t_{\alpha/2,\, n-1}$ or $t > t_{\alpha/2,\, n-1}$

Test statistic: $t = \dfrac{\overline{d}}{s_d/\sqrt{n}}$, where $\overline{d}$ is the mean of the differences between the two samples and s_d is the standard deviations of the differences; degrees of freedom $= n - 1$

Example

A Six Sigma trainer in a manufacturing facility intends to train 12 Six Sigma Green Belt candidates using an improved method of teaching Six Sigma. First the trainer administers a written test to the candidates to determine how well they know Six Sigma, then he conducts the training. The test is administered again after the retraining is completed. The following table contains the results.

Employee	Pre-training % correct	Post-training % correct	Difference
1	85	87	−2
2	81	93	−12
3	80	84	−4
4	77	84	−7
5	72	82	−10
6	70	76	−6
7	86	97	−11
8	83	91	−8
9	79	83	−4
10	89	92	−3
11	69	78	−9
12	82	90	−8

The mean of the pre-training results is 79.42 with a standard deviation of 6.37, and the mean of the post-training results is 86.42. The trainer wants to determine if the increase in test scores is statistically significant.

Procedure

Step 1: Set the null hypothesis. The question is whether or not the difference between the means of the two sets of results is 0, so a two-tailed hypothesis test is used.

$$H_0: \mu_1 - \mu_2 = 0$$

Step 2: Set the alternative hypothesis.

$$H_a: \mu_1 - \mu_2 \neq 0$$

Step 3: Set the alpha level and determine rejection criteria. We are using $\alpha = 0.05$ and must calculate the degrees of freedom, which is $n - 1$, so $12 - 1 = 11$. We then look at the t distribution table and determine the t value for a two-tailed test with 11 degrees of freedom and an alpha of 0.05. We find that the t value is 2.201.

Step 4: Determine the test statistic. The test statistic is $t = \dfrac{\bar{d}}{s_d/\sqrt{n}}$. The mean of the difference

$$\bar{d} = \frac{(-2) + (-12) + (-4) + (-7) + (-10) + (-6) + (-11) + (-8) + (-4) + (-3) + (-9) + (-8)}{12} =$$

$$\frac{-84}{12} = -7$$

The standard deviation of the difference s_d is

$$\sqrt{\frac{\Sigma(d - \bar{d})^2}{n - 1}} = 3.247$$

We enter these results into the test statistic $\dfrac{-7}{3.247/\sqrt{12}} = -7.468$.

Conclusion

The rejection region is lower than −2.201 or higher than 2.201; −7.47 is not within the acceptance region, so we reject the null hypothesis. The evidence supports the alternative hypothesis that there is a difference between the means.

Minitab Instructions

Step 1: Enter the data into a Minitab worksheet and go to "Stat > Basic Statistics > μ μ Paired t…"

Step 2: Select "Each sample is in a column."

Step 3: Click "Options…" and enter the desired confidence level and the alternative test. The default setting for confidence level is 95.0. Leave the alternative as "Difference ≠ hypothesized difference." Leave the hypothesized difference as 0.0.

Step 4: Click OK.

Minitab Output

Paired T-Test and CI: C1; C2
Paired T for C1 – C2

	N	Mean	StDev	SE Mean
C1	12	79.42	6.37	1.84
C2	12	86.42	6.35	1.83
Difference	12	−7.000	3.247	0.937

95% CI for mean difference: (−9.063; −4.937)

T-Test of mean difference = 0 (vs ≠ 0): T-Value = −7.47
P-Value = 0.000

Interpretation of Minitab Output

We use the *P* value calculated by Minitab in place of the *t* value calculated using the Student's *t* distribution. This Minitab-calculated *P* value of 0.000 is less than the alpha of 0.05, so we reject the null hypothesis in favor of the alternative hypothesis.

One-Tailed Paired t-*Test (Upper Tail)*

> Assumptions: Sample size $n \leq 30$, sample sizes are equal, population is normally distributed, standard deviations are approximately equal. Each population is tested twice, with observations paired together; the observations within the populations must be independent of each other.
>
> Null hypothesis (H_0): $\mu_d \leq \mu_0$
>
> Alternative hypothesis (H_a): $\mu_d > \mu_0$, with rejection criteria $t > t_{\alpha,\, n-1}$
>
> Test statistic: $t = \dfrac{\bar{d}}{s_d/\sqrt{n}}$, where $\bar{d}$ is the mean of the differences between the two samples and s_d is the standard deviations of the differences; degrees of freedom = $n - 1$

Example

Suppose the trainer from the previous example wants to know if the "before" scores are higher than the "after" scores. The mean of the difference $\bar{d} = -7$ and the standard deviation of the difference $s_d = 3.247$.

Procedure

Step 1: Set the null hypothesis. A one-tailed upper tail test is used to determine if the difference in means is greater than or equal to zero.

$$H_0:\ \mu_d \leq \mu_0$$

Step 2: Set the alternative hypothesis.

$$H_a:\ \mu_d > \mu_0$$

Step 3: Set the alpha level and determine rejection criteria. We are using $\alpha = 0.05$ and must calculate the degrees of freedom, which is $n - 1$, so $12 - 1 = 11$. We then look at the t distribution table and determine the t value for a one-tailed test with 11 degrees of freedom and an alpha of 0.05. We find that the t value is 1.796.

Step 4: Determine the test statistic.

$$t = \frac{-7}{3.247/\sqrt{12}} = -7.468$$

Conclusion

The rejection region is higher than 1.796, and −7.468 is less than 1.796, so we fail to reject the null hypothesis. There is insufficient evidence to support the alternative hypothesis that the difference in means is greater than 0.

Minitab Instructions

Step 1: Go to "Stat > Basic Statistics > μ μ Paired t…"

Step 2: Select "Summarized data (differences)."

Step 3: Enter the sample size, the mean of the difference ($\bar{d}$), and the standard deviation of the difference (s_d).

Step 4: Click "Options…" and enter the desired confidence level and the alternative test. The default setting for confidence level is 95.0. Change the alternative to "Difference > hypothesized difference." Leave the hypothesized difference as 0.0.

Step 5: Click OK.

Minitab Output

Paired T-Test and CI

	N	Mean	StDev	SE Mean
Difference	12	−7.000	3.247	0.937

95% lower bound for mean difference: −8.683

T-Test of mean difference = 0 (vs > 0): T-Value = −7.47
P-Value = 1.000

Interpretation of Minitab Output

We use the P value calculated by Minitab in place of the t value calculated using the Student's t distribution. This Minitab-calculated P value of 1.000 is greater than the alpha of 0.05, so we fail to reject the null hypothesis.

One-Tailed Paired t-Test (Lower Tail)

> Assumptions: Sample size $n \leq 30$, sample sizes are equal, population is normally distributed, standard deviations are approximately equal. Each population is tested twice, with observations paired together; the observations within the populations must be independent of each other.

Null hypothesis (H_0): $\mu_d \geq \mu_0$

Alternative hypothesis (H_a): $\mu_d < \mu_0$, with rejection criteria $t < t_{\alpha, n-1}$

Test statistic: $t = \dfrac{\overline{d}}{s_d/\sqrt{n}}$, where $\overline{d}$ is the mean of the differences between the two samples and s_d is the standard deviations of the differences; degrees of freedom $= n - 1$

Example

Suppose the trainer from the previous examples wants to know if the "before" scores are lower than the "after" scores. The mean of the difference $\overline{d} = -7$ and the standard deviation of the difference $s_d = 3.247$.

Procedure

Step 1: Set the null hypothesis. A one-tailed lower tail test is used to determine if the difference in means is less than zero.

$$H_0: \mu_d \geq \mu_0$$

Step 2: Set the alternative hypothesis.

$$H_a: \mu_d < \mu_0$$

Step 3: Set the alpha level and determine rejection criteria. We are using $\alpha = 0.05$ and must calculate the degrees of freedom, which is $n - 1$, so $12 - 1 = 1$. We then look at the t distribution table and determine the t value for a one-tailed test with 11 degrees of freedom and an alpha of 0.05. We find that the t value is 1.796, and this is a lower tail test, so it is a negative number.

Step 4: Determine the test statistic.

$$t = \frac{-7}{3.247/\sqrt{12}} = -7.468$$

Conclusion

The rejection region is lower than -1.796, and -7.468 is less than -1.796, so we reject the null hypothesis. The evidence supports the alternative hypothesis that the difference in means is less than zero.

Minitab Instructions

Step 1: Go to "Stat > Basic Statistics > μ μ Paired t…"

Step 2: Select "Summarized data (differences)."

Step 3: Enter the sample size, the mean of the difference ($\overline{d}$), and the standard deviation of the difference (s_d).

Step 4: Click "Options…" and enter the desired confidence level and the alternative test. The default setting for confidence level is 95.0. Change the alternative to "Difference < hypothesized difference." Leave the hypothesized difference as 0.0.

Step 5: Click OK.

Minitab Output

Paired T-Test and CI

	N	Mean	StDev	SE Mean
Difference	12	−7.000	3.247	0.937

95% upper bound for mean difference: −5.317

T-Test of mean difference = 0 (vs < 0): T-Value = −7.47
P-Value = 0.000

Interpretation of Minitab Output

We use the P value calculated by Minitab in place of the t value calculated using the Student's t distribution. This Minitab-calculated P value of 0.000 is less than the alpha of 0.05, so we reject the null hypothesis in favor of the alternative hypothesis.

Tests for Population Proportions

Hypothesis testing can also be performed for proportions. A proportion is found by dividing the number of occurrences by the number of opportunities for an occurrence. This can also be referred to as the number of successes or failures out of a total number of trials. Another way of phrasing proportions is p = number of cases / total population (Boslaugh and Watters 2008). Suppose there are 350 parts in a box and 14 are found to be defective; this results in a p of 0.04. A hypothesis test for a proportion assumes the distribution is binomial.

Hypothesis Tests for One-Sample Proportions

Two-Tailed Z-Test

Assumptions: Sample size $n > 30$, distribution is binomial, standard deviation is not relevant

Null hypothesis (H_0): $p = p_0$

Alternative hypothesis (H_a): $p \neq p_0$, with rejection criteria $Z < Z_{\alpha/2}$ or $Z > Z_{\alpha/2}$

Test statistic: $Z = \dfrac{(p - p_0)}{\sqrt{\dfrac{\hat{p}_0(1 - \hat{p}_0)}{n}}}$, where p is the sample proportion (occurrences/

opportunities) and p_0 is a hypothesized value for the proportion

Example

A Six Sigma Green Belt finds 14 defective parts out of 107 parts that were checked; this is a *p* of 0.1308 (14 defective parts / 107 parts checked). The Six Sigma Green Belt wants to know if the true population proportion defective is equal to 0.10.

Procedure

Step 1: Set the null hypothesis. We want to know if the actual proportion and the hypothesized proportion are equal, so we use a two-tailed test.

$$H_0: p = 0.10$$

Step 2: Set the alternative hypothesis.

$$H_a: p \neq 0.10$$

Step 3: Set the alpha level and determine rejection criteria. We are using $\alpha = 0.05$, so we use 1.96.

Step 4: Determine the test statistic.

$$Z = \frac{0.1308 - 0.10}{\sqrt{\dfrac{0.10(1 - 0.10)}{107}}} = 1.06$$

Conclusion

In this case the rejection region is less than −1.96 or greater than 1.96, and 1.06 is within the acceptance region, so we fail to reject the null hypothesis. There is insufficient evidence to support the alternative hypothesis that the true proportion defective is not equal to 10%.

Minitab Instructions

Step 1: Go to "Stat > Basic Statistics >1 Proportion…"

Step 2: Select "Summarized data" and enter the number of events and number of trials. *Note:* It is also possible to select entire columns to enter the data.

Step 3: Check the box next to "Perform hypothesis test." Enter the hypothesized proportion of 0.10.

Step 4: Click "Options…" and enter the desired confidence level and the alternative test. The default setting for confidence level is 95.0. Leave the alternative as "Proportion ≠ hypothesized proportion." Select "Normal approximation."

Step 5: Click OK.

Minitab Output

Test and CI for One Proportion
Test of p = 0.1 vs p ≠ 0.1

Sample	X	N	Sample p	95% CI	Z-Value	P-Value
1	14	107	0.130841	(0.066944; 0.194738)	1.06	0.288

Using the normal approximation.

Interpretation of Minitab Output

The Minitab-calculated P value of 0.288 is greater than the alpha of 0.05, so we fail to reject the null hypothesis.

One-Tailed Z-Test (Upper Tail)

Assumptions: Sample size $n > 30$, distribution is binomial, standard deviation is not relevant

Null hypothesis (H_0): $p \leq p_0$

Alternative hypothesis (H_a): $p > p_0$, with rejection criteria $Z > Z_\alpha$

Test statistic: $Z = \dfrac{(p - p_0)}{\sqrt{\dfrac{\hat{p}_0(1 - \hat{p}_0)}{n}}}$, where p is the sample proportion

(occurrences / opportunities) and p_0 is a hypothesized value for the proportion

Example

An inspector checks a shipment of 250 parts and finds 38 defective parts. The p is 0.152 (38 defective parts / 250 parts checked), and the inspector wants to be sure the true defective rate is greater than 25%.

Procedure

Step 1: Set the null hypothesis. We want to know if the proportion is greater than the hypothesized value, so we use a one-tailed upper tail test.

$$H_0: p \leq 0.25$$

Step 2: Set the alternative hypothesis.

$$H_a: p > 0.25$$

Step 3: Set the alpha level and determine rejection criteria. We are using $\alpha = 0.05$ and a one-tailed test, so we use 1.645.

Step 4: Determine the test statistic.

$$Z = \frac{0.152 - 0.25}{\sqrt{\dfrac{0.25(1 - 0.25)}{250}}} = -3.58$$

Conclusion

In this case the rejection region is greater than 1.645, and –3.58 is less than 1.645, so we fail to reject the null hypothesis. There is insufficient evidence to support the alternative hypothesis that the true proportion defective is greater than 25%.

Minitab Instructions

Step 1: Go to "Stat > Basic Statistics >1 Proportion..."

Step 2: Select "Summarized data" and enter the number of events and number of trials. *Note:* It is also possible to select entire columns to enter the data.

Step 3: Check the box next to "Perform hypothesis test." Enter the hypothesized proportion of 0.25.

Step 4: Click "Options..." and enter the desired confidence level and the alternative test. The default setting for confidence level is 95.0. Leave the alternative as "Proportion > hypothesized proportion." Select "Normal approximation."

Step 5: Click OK.

Minitab Output

Test and CI for One Proportion
Test of p = 0.25 vs p > 0.25

Sample	X	N	Sample p	95% Lower Bound	Z-Value	P-Value
1	38	250	0.152000	0.114651	–3.58	1.000

Using the normal approximation.

Interpretation of Minitab Output

The Minitab-calculated *P* value of 1.000 is greater than the alpha of 0.05, so we fail to reject the null hypothesis.

One-Tailed Z-Test (Lower Tail)

Assumptions: Sample size $n > 30$, distribution is binomial, standard deviation is not relevant

Null hypothesis (H_0): $p \geq p_0$

Alternative hypothesis (H_a): $p < p_0$, with rejection criteria $Z < Z_\alpha$

Test statistic: $Z = \dfrac{(p - p_0)}{\sqrt{\dfrac{\hat{p}_0(1 - \hat{p}_0)}{n}}}$, where p is the sample proportion

(occurrences / opportunities) and p_0 is a hypothesized value for the proportion

Example

An account manager randomly selects 86 invoices and determines 8 are overdue. The account manager wants to determine if the proportion overdue is equal to or less than 10%. The p is 0.093 (8 overdue / 86 invoices) and p_c is 0.10.

Procedure

Step 1: Set the null hypothesis. We want to know if the proportion is less than the hypothesized value, so we use a one-tailed lower tail test.

$$H_0: p \geq 0.10$$

Step 2: Set the alternative hypothesis.

$$H_a: p < 0.10$$

Step 3: Set the alpha level and determine rejection criteria. We are using $\alpha = 0.05$ and a one-tailed test, so we use 1.645.

Step 4: Determine the test statistic.

$$Z = \frac{0.093 - 0.10}{\sqrt{\dfrac{0.10(1 - 0.10)}{86}}} = -0.22$$

Conclusion

In this case the rejection region is less than −1.645, and −0.22 is greater than −1.645, so we fail to reject the null hypothesis. There is inducement evidence to support the alternative hypothesis that the true difference in proportions is less than 10%.

Minitab Instructions

Step 1: Go to "Stat > Basic Statistics > 1 Proportion…"

Step 2: Select "Summarized data" and enter the number of events and number of trials. *Note:* It is also possible to select entire columns to enter the data.

Step 3: Check the box next to "Perform hypothesis test." Enter the hypothesized proportion of 0.10.

Step 4: Click "Options…" and enter the desired confidence level and the alternative test. The default setting for confidence level is 95.0. Leave the alternative as "Proportion < hypothesized proportion." Select "Normal approximation."

Step 5: Click OK.

Minitab Output

Test and CI for One Proportion
Test of p = 0.1 vs p < 0.1

Sample	X	N	Sample p	95% Upper Bound	Z-Value	P-Value
1	8	86	0.093023	0.144543	−0.22	0.415

Using the normal approximation.

Interpretation of Minitab Output

The Minitab-calculated P value of 0.415 is greater than the alpha of 0.05, so we fail to reject the null hypothesis.

Hypothesis Tests for Two-Sample Proportions

Two-Tailed Z-Test

Assumptions: Sample size $np \geq 5$ and $nq \geq 5$, distribution is binomial, standard deviation is not relevant. The proportion (p) is equal to number of occurrences / number of opportunities; q is the chance that an event will not occur and can be determined using $1 - p = q$.

Null hypothesis (H_0): $p_1 - p_2 = 0$

Alternative hypothesis (H_a): $p_1 - p_2 \neq 0$, with rejection criteria $Z < Z_{\alpha/2}$ or $Z > Z_{\alpha/2}$

Test statistic: $Z = \dfrac{(p_1 - p_2)}{\sqrt{\hat{p}(1 - \hat{p})\left(\dfrac{1}{n_1} + \dfrac{1}{n_2}\right)}}$, with $\hat{p} = \dfrac{n_1 p_1 + n_2 p_2}{n_1 + n_2}$, where p_1 is the

proportion for sample one and p_2 is the proportion for sample two

Example

A Six Sigma Black Belt takes a sample from each of two production machines and determines that 9 out of 47 samples from the first machine are defective and 7 out of 42 samples from the second machine are defective. The Six Sigma Black Belt wants to know if the proportions defective from each machine are equal.

Procedure

Step 1: Set the null hypothesis. We want to know if the two populations are equal, so we use a two-tailed test.

$$H_0: p_1 - p_2 = 0$$

Step 2: Set the alternative hypothesis.

$$H_a: p_1 - p_2 \neq 0$$

Step 3: Set the alpha level and determine rejection criteria. We are using $\alpha = 0.05$, so we use 1.96.

Step 4: Determine the test statistic. The first step is calculating the weighted mean proportion used in the computation ($\hat{p}$), which is equal to:

$$\frac{(47)(0.191) + (42)(0.167)}{47 + 42} = 0.180$$

Then determine the Z score.

$$Z = \frac{0.191 - 0.167}{\sqrt{0.180(1 - 0.180)\left(\frac{1}{47} + \frac{1}{42}\right)}} = 0.30$$

Conclusion

In this case the rejection region is less than −1.96 or greater than 1.96, and 0.30 is within the acceptance region, so we fail to reject the null hypothesis. The evidence does not support the alternative hypothesis that there is a difference between the proportions.

Minitab Instructions

Step 1: Go to "Stat > Basic Statistics > 2 Proportions…"

Step 2: Select "Summarized data" and enter the number of events and number of trials for both samples.

Step 3: Click "Options…" and enter the desired confidence level and the hypothesized difference. For a two-tailed test use 0. Leave the alternative as "Proportion ≠ hypothesized proportion." Select "Use the pooled estimate of the proportion."

Step 4: Click OK.

Minitab Output

Test and CI for Two Proportions

Sample	X	N	Sample p
1	9	47	0.191489
2	7	42	0.166667

Difference = p (1) − p (2)

Estimate for difference: 0.0248227

95% CI for difference: (−0.134417; 0.184062)

Test for difference = 0 (vs ≠ 0): Z = 0.30
P-Value = 0.761

Fisher's exact test: P-Value = 0.790

Interpretation of Minitab Output

We use the P value calculated by Minitab in place of the Z value calculated using the Z-test. This Minitab-calculated P value of 0.761 is greater than the alpha of 0.05, so we fail to reject the null hypothesis in favor of the alternative hypothesis.

One-Tailed Z-Test (Upper Tail)

Assumptions: Sample size $np \geq 5$ and $nq \geq 5$, distribution is binomial, standard deviation is not relevant. The proportion (p) is equal to number of occurrences / number of opportunities; q is the chance that an event will not occur and can be determined using $1 - p = q$.

Null hypothesis (H_0): $p_1 - p_2 \leq 0$

Alternative hypothesis (H_a): $p_1 - p_2 > 0$, with rejection criteria $Z < Z_{\alpha/2}$ or $Z > Z_{\alpha/2}$

Test statistic: $Z = \dfrac{(p_1 - p_2)}{\sqrt{\hat{p}(1-\hat{p})\left(\frac{1}{n_1}+\frac{1}{n_2}\right)}}$, with $\hat{p} = \dfrac{n_1 p_1 + n_2 p_2}{n_1 + n_2}$, where p_1 is the

proportion for sample one and p_2 is the proportion for sample two

Example

The Six Sigma Black Belt from the previous example wants to determine if p_1 is greater than p_2. There are 9 out of 47 defective samples from machine one and 7 out of 42 defective samples from machine two.

Procedure

Step 1: Set the null hypothesis. We want to know if the proportion from population one is greater than the proportion from population two, so we use a one-tailed upper tail test.

$$H_0\text{: } p_1 - p_2 \leq 0$$

Step 2: Set the alternative hypothesis.

$$H_a\text{: } p_1 - p_2 > 0$$

Step 3: Set the alpha level and determine rejection criteria. We are using $\alpha = 0.05$ and a one-tailed test, so we use 1.645.

Step 4: Determine the test statistic. The first step is calculating $\hat{p}$, which is equal to:

$$\frac{(47)(0.191) + (42)(0.167)}{47 + 42} = 0.180$$

Then determine the Z score.

$$Z = \frac{0.191 - 0.167}{\sqrt{0.180(1 - 0.180)\left(\frac{1}{47} + \frac{1}{42}\right)}} = 0.30$$

Conclusion

In this case the rejection region is greater than 1.645, and 0.30 is less than 1.645, so we fail to reject the null hypothesis. The evidence does not support the alternative hypothesis that p_1 is greater than p_2.

Minitab Instructions

Step 1: Go to "Stat > Basic Statistics > 2 Proportions…"

Step 2: Select "Summarized data" and enter the number of events and number of trials for both samples.

Step 3: Click "Options…" and enter the desired confidence level and the hypothesized difference. For the alternative, select "Proportion > hypothesized proportion." Select "Use the pooled estimate of the proportion."

Step 4: Click OK.

Minitab Output

Test and CI for Two Proportions

Sample	X	N	Sample p
1	9	47	0.191489
2	7	42	0.166667

Difference = p (1) − p (2)

Estimate for difference: 0.0248227

95% upper bound for difference: 0.158461

Test for difference = 0 (vs < 0): Z = 0.30
P-Value = 0.620

Fisher's exact test: P-Value = 0.718

Interpretation of Minitab Output

We use the P value calculated by Minitab in place of the Z value calculated using the Z-test. This Minitab-calculated P value of 0.620 is greater than the alpha of 0.05, so we fail to reject the null hypothesis.

One-Tailed Z-Test (Lower Tail)

Assumptions: Sample size $np \geq 5$ and $nq \geq 5$, distribution is binomial, standard deviation is not relevant. The proportion (p) is equal to number of occurrences / number of opportunities; q is the chance that an event will not occur and can be determined using $1 - p = q$.

Null hypothesis (H_0): $p_1 - p_2 \geq 0$

Alternative hypothesis (H_a): $p_1 - p_2 < 0$, with rejection criteria $Z < Z_\alpha$

Test statistic: $Z = \dfrac{(p_1 - p_2)}{\sqrt{\hat{p}(1 - \hat{p})\left(\dfrac{1}{n_1} + \dfrac{1}{n_2}\right)}}$, with $\hat{p} = \dfrac{n_1 p_1 + n_2 p_2}{n_1 + n_2}$, where p_1 is the

proportion for sample one and p_2 is the proportion for sample two

Example

Suppose the Six Sigma Black Belt from the previous example wants to determine if p_1 is less than p_2. There are 9 out of 47 defective samples from machine one and 7 out of 42 defective samples from machine two.

Procedure

Step 1: Set the null hypothesis. We want to know if the proportion from population one is less than the proportion from population two, so we use a one-tailed lower tail test.

$$H_0: p_1 - p_2 \geq 0$$

Step 2: Set the alternative hypothesis.

$$H_a: p_1 - p_2 < 0$$

Step 3: Set the alpha level and determine rejection criteria. We are using $\alpha = 0.05$ and a one-tailed test, so we use 1.645.

Step 4: Determine the test statistic. The first step is calculating $\hat{p}$, which is equal to:

$$\frac{(47)(0.191) + (42)(0.167)}{47 + 42} = 0.180$$

Then determine the Z score.

$$Z = \frac{0.191 - 0.167}{\sqrt{0.180(1 - 0.180)\left(\frac{1}{47} + \frac{1}{42}\right)}} = 0.30$$

Conclusion

In this case the rejection region is less than −1.645, and 0.30 is greater than −1.645, so we fail to reject the null hypothesis. The evidence does not support the alternative hypothesis that p_1 is less than p_2.

Minitab Instructions

Step 1: Go to "Stat > Basic Statistics > 2 Proportions…"

Step 2: Select "Summarized data" and enter the number of events and number of trials for both samples.

Step 3: Click "Options…" and enter the desired confidence level and the hypothesized difference. For the alternative, select "Proportion < hypothesized proportion." Select "Use the pooled estimate of the proportion."

Step 4: Click OK.

Minitab Output

Test and CI for Two Proportions

Sample	X	N	Sample p
1	9	47	0.191489
2	7	42	0.166667

Difference = p (1) – p (2)

Estimate for difference: 0.0248227

95% lower bound for difference: –0.108815

Test for difference = 0 (vs > 0): Z = 0.30
P-Value = 0.380

Fisher's exact test: P-Value = 0.490

Interpretation of Minitab Output

We use the *P* value calculated by Minitab in place of the *Z* value calculated using the *Z*-test. This Minitab-calculated *P* value of 0.380 is greater than the alpha of 0.05, so we fail to reject the null hypothesis.

CONFIDENCE INTERVALS FOR MEANS

A confidence interval is an interval estimate of a population parameter that is used to determine the precision of an estimate (Metcalfe 1994). Our statistical results are only an estimate, and the confidence interval tells us where the true value lies within the selected degree of confidence. A confidence level is an interval in which the value of the statistic being considered is believed to fall. A confidence interval uses a confidence level selected by the experimenter; for example, a 95% confidence level is often used. The confidence level uses the Z score, with a 95% confidence level equal to 1.96 for a two-tailed test and 1.64 for a one-tailed test.

Sample Size Determination

A sample size can be determined that will result in a specified margin of error. The formula is

$$n = \left(\frac{Z * \sigma}{E}\right)^2$$

where *E* is the specified margin of error. This formula can be used to decide what sample size to use to ensure that the level of error is kept to an acceptable level.

Example

A designer wants to test the pull-off force in a prototype design and would like to have a confidence level of 95% (*Z* = 1.96). The prototype samples have a standard deviation of 33 and the designer wants a confidence interval that has a margin of error of 12.

Procedure

Step 1: Plug the numbers into the formula and calculate.

$$n = \left(\frac{Z * \sigma}{E}\right)^2 = \left(\frac{1.96 * 33}{12}\right)^2$$

Conclusion

The result is 29.05, so the designer will need to sample 30 parts to ensure a 95% confidence level.

Minitab Instructions

Step 1: Go to "Stat > Power and Sample Sizes > Sample Size for Estimation."

Step 2: Go to "Parameter" and select "Mean (Normal)."

Step 3: Under "Planning value standard deviation," enter the standard deviation.

Step 4: Select "Estimate sample sizes."

Step 5: Enter the margin of error under "Margins of error for confidence intervals."

Step 6: Select "Options" to select a confidence level and to choose between a one-tailed lower tail test, a two-tailed test, and a one-tailed upper tail test. Check the box next to "Assume population standard deviation is known."

Step 7: Click OK.

Minitab Output

Sample Size for Estimation

Method	
Parameter	Mean
Distribution	Normal
Standard deviation	33 (population value)
Confidence level	95%
Confidence interval	Two-sided

Results	
Margin of Error	Sample Size
12	30

Interpretation of Minitab Output

Minitab has determined that 30 samples will be needed to have a margin of error of 12.

Confidence Interval for the Mean with Known Standard Deviation

Assumptions: Sample size $n \geq 30$, any population type, standard deviation is known; or sample size $n < 30$, population is normal, standard deviation is known

Test statistic: $\bar{x} \pm Z_{\alpha/2} * \dfrac{\sigma}{\sqrt{n}}$; degrees of freedom = n

Example

In an earlier example, an engineer determined that a sample of 32 parts with an average fatigue life of 674 hours was from a population with a mean of 678 hours and a standard deviation of 18. The engineer now wants to construct a confidence interval for the results.

Procedure

Step 1: Select the confidence level. In this case, the engineer wants 95% confidence.

Step 2: Enter the data into the test statistic.

$$674 - 1.96 * \frac{18}{\sqrt{32}} = 667.76 \text{ and } 674 + 1.96 * \frac{18}{\sqrt{32}} = 680.24$$

Conclusion

Interpret the results. The engineer can be 95% confident that the true mean lies somewhere between 667.76 and 680.24.

Minitab Instructions

See instructions for a Z-test with standard deviation known. The confidence interval for the mean is provided by Minitab along with the results for the hypothesis test.

Minitab Output

One-Sample Z
Test of mu = 678 vs not = 678
The assumed standard deviation = 18

N	Mean	SE Mean	95% CI	Z	P
32	674.00	3.18	(667.76; 680.24)	−1.26	0.209

Interpretation of Minitab Output

Minitab automatically calculates a confidence interval when performing the Z-test, so these results are exactly the same as in the previous example. The confidence interval given by Minitab is 667.76 for the lower confidence limit and 680.24 for the upper confidence limit.

Confidence Interval for the Mean with Unknown Standard Deviation

Assumptions: Sample size $n \geq 30$, any population type, population standard deviation is unknown

Test statistic: $\bar{x} \pm Z_{\alpha/2} * \dfrac{s}{\sqrt{n}}$

Example

A quality engineer wants a 95% ($Z = 1.96$) confidence interval for a sample size of 57 parts with a mean of 24 mm and a population standard deviation of 4.3.

Procedure

Step 1: Plug the numbers into the test statistic.

$$24 - 1.96 * \frac{4.3}{\sqrt{57}} = 22.9 \text{ and } 24 + 1.96 * \frac{4.3}{\sqrt{57}} = 25.1$$

Conclusion

Interpret the results. The 95% confidence interval for the mean is between 22.9 and 25.1.

Minitab Instructions

See instructions for a *t*-test with standard deviation unknown. The confidence interval for the mean is provided by Minitab along with the results for the hypothesis test.

Minitab Output

One-Sample Z
The assumed standard deviation = 4.3

N	Mean	SE Mean	95% CI
57	24.000	0.570	(22.884; 25.116)

Interpretation of Minitab Output

Minitab has determined that the 95% confidence interval is from 22.884 to 25.116.

Confidence Interval for the Mean with Unknown Standard Deviation and Small Sample Size

Assumptions: Sample size $n < 30$, population distribution is normal, population standard deviation is unknown

Test statistic: $\bar{x} \pm t_{\alpha/2} * \frac{s}{\sqrt{n}}$; degrees of freedom = $n - 1$

Example

An engineer has measured the length of 17 parts and found the mean to be 1435 mm, with a sample standard deviation of 32. What is the 95% confidence interval for the mean?

Procedure

Step 1: Determine the degrees of freedom ($17 - 1 = 16$) and find the t value for the appropriate confidence level. In this case the alpha is set at 0.05, and this is 2.131 in the Student's t distribution with an alpha of 0.05 using the two-tailed table.

Step 2: Plug the numbers into the test statistic.

$$1435 - 2.131 * \frac{32}{\sqrt{17}} = 1418.5 \text{ and } 1435 + 2.131 * \frac{32}{\sqrt{17}} = 1451.5$$

Conclusion

Interpret the results. The 95% confidence interval for the mean is between 1418.5 and 1451.5.

Minitab Instructions

See instructions for a t-test with standard deviation unknown. The confidence interval for the mean is provided by Minitab along with the results for the hypothesis test.

Minitab Output

One-Sample T

N	Mean	StdDev	SE Mean	95% CI
17	1435.00	32.00	7.76	(1418.55; 1451.45)

Interpretation of Minitab Output

Minitab has determined that the 95% confidence interval is from 1418.55 to 1451.45.

Sample Size Proportion Tests

Sample size determination: $n = \left(\dfrac{Z_{\alpha/2}}{E}\right)^2 \hat{p}(1 - \hat{p})$, where E is the margin of error in percent and $\hat{p}$ is the estimated proportion.

Example

An engineer will be sampling the proportion of components that are defective and is willing to accept a margin of error (E) of 8% and wants 95% confidence ($Z = 1.96$). The estimated proportion is 20%. How many samples will be needed?

Procedure

Step 1: Plug the numbers into the formula.

$$n = \left(\frac{1.96}{0.08}\right)^2 * 0.20(1 - 0.20) = 96.04$$

Conclusion

Round up to the next whole number; therefore, a sample size of 97 will be needed.

Minitab Instructions

Step 1: Select "Stat > Power and Sample Size > Sample Size for Estimation…"

Step 2: Go to "Parameter" and select "Proportion (Binomial)."

Step 3: Under "Planning Value Proportion," enter the estimated proportion.

Step 4: Select "Estimate sample sizes."

Step 5: Enter the margin of error under "Margins of error for confidence intervals."

Step 6: Select "Options" to select a confidence level and to choose between a one-tailed lower tail test, a two-tailed test, and a one-tailed upper tail test.

Minitab Output

Sample Size for Estimation	
Method	
Parameter	Proportion
Distribution	Binomial
Proportion	0.2
Confidence level	95%
Confidence interval	Two-sided
Results	
Margin of Error	Sample Size
0.08	128

Interpretation of Minitab Output

Minitab has determined that a sample size of 128 will be needed. This difference in results is due to Minitab's use of a method based on the F statistic, which is more sophisticated than the normal approximation method based on the Z statistic.

Confidence Interval for Proportions

Assumptions: Sample size $n \geq 30$, distribution is binomial, standard deviation is not relevant

Confidence interval: $p \pm Z_{\alpha/2} * \sqrt{\dfrac{\hat{p}(1-\hat{p})}{n}}$

Example

A Six Sigma Black Belt samples 842 parts and determines that 34 are defective. What is the confidence interval for the proportion?

Procedure

Step 1: Determine the value for p. The value p is the number of occurrences divided by the number of opportunities, so $p = 34 / 842 = 0.040$.

Step 2: Determine the value for Z. Using 95% confidence, the Z value would be 1.96.

Step 3: Plug the numbers into the formula.

$$0.040 - 1.96 * \sqrt{\frac{0.040(1 - 0.040)}{842}} = 0.026 \text{ and } 0.040 + 1.96 * \sqrt{\frac{0.040(1 - 0.040)}{842}} = 0.538$$

Conclusion

Interpret the results. The 95% confidence interval is from 0.026 to 0.538.

Minitab Instructions

See instructions for a hypothesis test with a proportion. The confidence interval for the mean is provided by Minitab along with the results for the hypothesis test.

Minitab Output

Test and CI for One Proportion

Sample	X	N	Sample p	95% CI
1	34	842	0.040380	(0.028124; 0.055972)

Interpretation of Minitab Output

The 95% confidence interval calculated by Minitab is from 0.028 to 0.056. This varies slightly from the manual results because Minitab uses more significant digits in the calculations.

CHI-SQUARE TEST OF POPULATION VARIANCE

The chi-square test, also written as "χ^2 test," is used to compare a sample variance to a hypothetical population variance using the χ^2 (pronounced "kye square") sampling distribution. The degrees of freedom for this test are equal to $n - 1$. Like the t distribution, the chi-square distribution approaches the shape of the normal distribution as the sample size increases (McClave and Dietrich 1991); however, a chi-square distribution is always left limited at zero, and there can never be a negative chi-square value. Figure 3.7 shows a chi-square distribution with 2 degrees of freedom and 5 degrees of freedom. Figure 3.8 shows a chi-square distribution with 10 degrees of freedom and 20 degrees of freedom. The chi-square distribution approaches the normal distribution when the number of degrees of freedom is increased. However, it will not match the normal distribution.

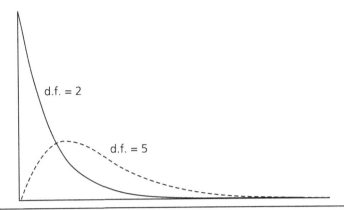

Figure 3.7   The chi-square sampling distribution with 2 degrees of freedom and 5 degrees of freedom.

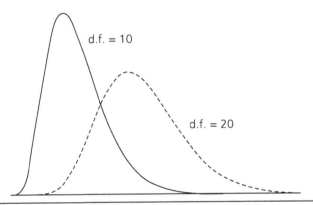

Figure 3.8   The chi-square sampling distribution with 10 degrees of freedom and 20 degrees of freedom.

The chi-square distribution is not symmetrical like the Z distribution. Using an alpha level of 0.05 for a two-tailed test requires looking in the chi-square table for 0.025 for the lower tail and 0.975 for the upper tail. A one-tailed upper tail test with an alpha of 0.05 requires looking up a cumulative probability of 0.95, and a one-tailed lower tail test with an alpha of 0.05 requires a cumulative probability of 0.05.

Two-Tailed Chi-Square Test

Assumptions: Population

Null hypothesis (H_0): $\sigma = \sigma_0$

Alternative hypothesis (H_a): $\sigma \neq \sigma_0$, with rejection criteria $\chi^2 < \chi^2_{n-1,\,1-\alpha/2}$ or $\chi^2 >_{n-1,\,1-\alpha/2}$

Test statistic: $\chi^2 = (n-1)\dfrac{s^2}{\sigma_0^2}$; degrees of freedom $= n-1$

Example

A Six Sigma Black Belt takes a sample size of 9 from a process and determines that the sample standard deviation is 38. The process standard deviation is believed to be 42.

Procedure

Step 1: Set the null hypothesis.

$$H_0: \sigma = 42$$

Step 2: Set the alternative hypothesis.

$$H_a: \sigma \neq 42$$

Step 3: Set the alpha level and determine rejection criteria. The alpha level is 0.05. This means we use 0.025 for the lower tail and 0.975 for the upper tail. With degrees of freedom equal to 9 – 1, we look at the χ^2 distribution table and determine that the value of 8 degrees of freedom is 2.70 for the lower tail and 19.0 for the upper tail.

Step 4: Determine the test statistic.

$$\chi^2 = (9-1)\frac{38^2}{42^2} = 6.55$$

Conclusion

The resulting test statistic is between the critical values of 2.70 and 19.0, so we fail to reject the null hypothesis. The evidence does not support the alternative hypothesis that the standard deviation is not equal to 42.

Minitab Instructions

Step 1: Go to "Stat > Basic Statistics > σ^2 1 Variance…"

Step 2: Select "Sample standard deviation" and enter the sample size and sample standard deviation. *Note:* It is also possible to select entire columns to enter the data. Check the box next to "Perform hypothesis test" and enter the hypothesized standard deviation.

Step 3: Click "Options…" and enter the desired confidence level and the alternative test. The default setting for confidence level is 95.0. Change the alternative to "Standard deviation ≠ hypothesized standard deviation."

Step 4: Click OK.

Minitab Output

Test and CI for One Variance

Method

| Null hypothesis | $\sigma = 42$ |
| Alternative hypothesis | $\sigma \neq 42$ |

The chi-square method is only for the normal distribution.
The Bonett method cannot be calculated with summarized data.

Statistics

N	StDev	Variance
9	38.0	1444

95% Confidence Intervals

Method	CI for StDev	CI for Variance
Chi-Square	(25.7; 72.8)	(659; 5300)

Test

Method	Test Statistic	DF	P-Value
Chi-Square	6.55	8	0.828

Interpretation of Minitab Output

The Minitab *P* value is greater than the alpha of 0.05, so we fail to reject the null hypothesis.

One-Tailed Chi-Square Test (Upper Tail)

Assumptions: Population

Null hypothesis (H_0): $\sigma \leq \sigma_0$

Alternative hypothesis (H_a): $\sigma > \sigma_0$, with rejection criteria $\chi^2 > \chi^2_{n-1,\ \alpha}$

Test statistic: $\chi^2 = (n-1) \dfrac{s^2}{\sigma_0^2}$; degrees of freedom $= n - 1$

Example

A Six Sigma Green Belt determines that the standard deviation of a sample size of 17 is 239, and the population standard deviation is believed to be 144. The Six Sigma Green Belt needs to determine if the population standard deviation is actually higher than it is believed to be.

Procedure

Step 1: Set the null hypothesis.

$$H_0: \sigma \leq 144$$

Step 2: Set the alternative hypothesis.

$$H_a: \sigma > 144$$

Step 3: Set the alpha level and determine rejection criteria. The alpha level is 0.05, so we use 0.95. With degrees of freedom equal to 17 – 1, we look at the χ^2 distribution table and find a value of 16 when the degrees of freedom is 26.3 for a one-tailed upper tail test.

Step 4: Determine the test statistic.

$$\chi^2 = (17 - 1)\frac{239^2}{144^2} = 44.07$$

Conclusion

The resulting test statistic is greater than 26.3, so we reject the null hypothesis. The evidence supports the alternative hypothesis that the population standard deviation is greater than 144.

Minitab Instructions

Step 1: Go to "Stat > Basic Statistics > σ^2 1 Variance…"

Step 2: Select "Sample standard deviation" and enter the sample size and sample standard deviation. *Note:* It is also possible to select entire columns to enter the data. Check the box next to "Perform hypothesis test" and enter the hypothesized standard deviation.

Step 3: Click "Options…" and enter the desired confidence level and the alternative test. The default setting for confidence level is 95.0. Change the alternative to "Standard deviation > hypothesized standard deviation."

Step 4: Click OK.

Minitab Output

Test and CI for One Variance

Method		
Null hypothesis	$\sigma = 144$	
Alternative hypothesis	$\sigma > 144$	

The chi-square method is only for the normal distribution.

The Bonett method cannot be calculated with summarized data.

Statistics

N	StDev	Variance
17	239	57121

95% Confidence Intervals

Method	Lower Bound for StDev	Lower Bound for Variance
Chi-Square	186	34755

Test

Method	Test Statistic	DF	P-Value
Chi-Square	44.07	16	0.000

Interpretation of Minitab Output

The Minitab-generated *P* value is less than the alpha value of 0.05, so we reject the null hypothesis in favor of the alternative hypothesis.

One-Tailed Chi-Square Test (Lower Tail)

Assumptions: Population

Null hypothesis (H_0): $\sigma \geq \sigma_0$

Alternative hypothesis (H_a): $\sigma < \sigma_0$, with rejection criteria $\chi^2 < \chi^2_{n-1,\,1-\alpha}$

Test statistic: $\chi^2 = (n-1)\dfrac{s^2}{\sigma_0^2}$; degrees of freedom $= n - 1$

Example

A Six Sigma project team member determines that the standard deviation of a sample of 14 is 44, and the population standard deviation is believed to be 49. The team member needs to determine if the population standard deviation is lower than it is believed to be.

Procedure

Step 1: Set the null hypothesis.

$$H_0: \sigma \geq 49$$

Step 2: Set the alternative hypothesis.

$$H_a: \sigma < 49$$

Step 3: Set the alpha level and determine rejection criteria. The alpha level is 0.05. With degrees of freedom equal to 14 − 1, we look at the χ^2 distribution table and find that the value of 13 degrees of freedom is 5.89.

Step 4: Determine the test statistic.

$$\chi^2 = (14 - 1)\frac{44^2}{49^2} = 10.49$$

Conclusion

The resulting test statistic is greater than 5.89, so we fail to reject the null hypothesis. The evidence does not support the alternative hypothesis that the population standard deviation is less than 49.

Minitab Instructions

Step 1: Go to "Stat > Basic Statistics > σ^2 1 Variance…"

Step 2: Select "Sample standard deviation" and enter the sample size and sample standard deviation. *Note:* It is also possible to select entire columns to enter the data. Check the box next to "Perform hypothesis test" and enter the hypothesized standard deviation.

Step 3: Click "Options..." and enter the desired confidence level and the alternative test. The default setting for confidence level is 95.0. Change the alternative to "Standard deviation < hypothesized standard deviation."

Step 4: Click OK.

Minitab Output

Test and CI for One Variance

Method	
Null hypothesis	$\sigma = 49$
Alternative hypothesis	$\sigma < 49$

The chi-square method is only for the normal distribution.
The Bonett method cannot be calculated with summarized data.

Statistics

N	StDev	Variance
14	44.0	1936

95% One-Sided Confidence Intervals

Method	Upper Bound for StDev	Upper Bound for Variance
Chi-Square	65.4	4272

Test

Method	Test Statistic	DF	P-Value
Chi-Square	10.48	13	0.346

Interpretation of Minitab Output

The Minitab P value is greater than the alpha of 0.05, so we fail to reject the null hypothesis.

F-TEST FOR THE VARIANCE OF TWO SAMPLES

There are occasions when it is necessary to compare variances and not means, such as when analyzing the precision of measuring devices (McClave and Dietrich 1991). On such occasions the F-test is the appropriate test. The F-test uses the variances of samples taken from two populations with a normal distribution.

Two-Tailed F-Test

Assumptions: Both populations are normally distributed

Null hypothesis (H_0): $\sigma_1^2 = \sigma_2^2$

Alternative hypothesis (H_a): $\sigma_1^2 \neq \sigma_2^2$, with rejection criteria $F > F_{\alpha/2,\, n1-1,\, n2-2}$ or $F < F_{1-\alpha/2,\, n1-1,\, n2-1}$

Test statistic: $F = \dfrac{s_1^2}{s_2^2}$; degrees of freedom = $n_1 - 1, n_2 - 1$

Example

A quality engineer wants to determine if the variances of two production machines are comparable. A sample of 15 is taken from the first machine and found to have a standard deviation of 0.63, and a sample of 17 is taken from the second machine and found to have a standard deviation of 0.49.

Procedure

Step 1: Set the null hypothesis.

$$H_0: \sigma_1^2 = \sigma_2^2$$

Step 2: Set the alternative hypothesis.

$$H_a: \sigma_1^2 \neq \sigma_2^2$$

Step 3: Set the alpha level and determine rejection criteria. The alpha level is 0.05. With degrees of freedom equal to 15 − 1 = 14 for the numerator and 17 − 1 = 16 for the denominator, we look at the *F* distribution table and determine that the critical value is 2.37.

Step 4: Determine the test statistic.

$$F = \frac{0.63^2}{0.49^2} = 1.653$$

Conclusion

The test statistic of 1.653 is less than 2.37, so we fail to reject the null hypothesis. The evidence does not support the alternative hypothesis that the variances are not comparable.

Minitab Instructions

Step 1: Go to "Stat > Basic Statistics > σ2/σ2 Variances…"

Step 2: Select "Sample variances."

Step 3: Enter the sample size and standard deviation for sample one and sample two.

Step 4: Click "Options" to enter the confidence level. The default setting is 95.0.

Step 5: Under "Ratio," select "(sample 1 standard deviation) / (sample 2 standard deviation)" for standard deviations, or "(Sample variance 1 / Sample variance 2)" if variance is being used.

Step 6: Under "Alternative hypothesis," select "ratio ≠ hypothesized ratio" for a two-tailed test.

Step 7: Click OK.

Minitab Output

Test and CI for Two Variances

Method

Null hypothesis	σ(First) / σ(Second) = 1
Alternative hypothesis	σ(First) / σ(Second) ≠ 1
Significance level	α = 0.05

F method was used. This method is accurate for normal data only.

Statistics

Sample	N	StDev	Variance	95% CI for StDevs
First	15	0.630	0.397	(0.461; 0.994)
Second	17	0.490	0.240	(0.365; 0.746)

Ratio of standard deviations = 1.286
Ratio of variances = 1.653

95% Confidence Intervals

Method	CI for StDev Ratio	CI for Variance Ratio
F	(0.766; 2.198)	(0.587; 4.833)

Test

Method	DF1	DF2	Test Statistic	P-Value
F	14	16	1.65	0.334

Interpretation of Minitab Output

The Minitab-generated *P* value is greater than the alpha of 0.05, so we fail to reject the null hypothesis.

One-Tailed *F*-Test (Upper Tail)

Assumptions: Both populations are normally distributed

Null hypothesis (H_0): H_0: $\sigma_1^2 \leq \sigma_2^2$

Alternative hypothesis (H_a): H_a: $\sigma_1^2 > \sigma_2^2$, with rejection criteria $F > F_{\alpha,\, n1-1,\, n2-1}$

Test statistic: $F = \dfrac{s_1^2}{s_2^2}$; degrees of freedom = $n_1 - 1, n_2 - 1$

Example

Two measuring devices are checked for precision. The first one has a standard deviation of 0.094 with 12 samples, and the second one has a sample size of 18 and a standard deviation of 0.054. Is the variability of the first device really greater than the variability of the second device?

Procedure

Step 1: Set the null hypothesis.

$$H_0: \sigma_1^2 \le \sigma_2^2$$

Step 2: Set the alternative hypothesis.

$$H_a: \sigma_1^2 > \sigma_2^2$$

Step 3: Set the alpha level and determine rejection criteria. The alpha level is 0.05. With degrees of freedom equal to 12 − 1 = 11 for the numerator and 18 − 1 = 17 for the denominator, we look at the F distribution table and determine that the critical value is 2.41.

Step 4: Determine the test statistic.

$$F = \frac{0.094^2}{0.054^2} = 3.030$$

Conclusion

The test statistic of 3.030 is greater than the critical value of 2.41, so we reject the null hypothesis. The evidence supports the alternative hypothesis that the variance of the first measuring device is greater than the variance of the second measuring device.

Minitab Instructions

Step 1: Go to "Stat > Basic Statistics > σ^2/σ^2 Variances…"

Step 2: Select "Sample variances."

Step 3: Enter the sample size and standard deviation for sample one and sample two.

Step 4: Click "Options."

Step 5: Click "Options" to enter the confidence level. The default setting is 95.0.

Step 6: Under "Ratio," select "(sample 1 standard deviation) / (sample 2 standard deviation)" for standard deviations, or "(Sample variance 1 / Sample variance 2)" if variance is being used.

Step 7: Under "Alternative hypothesis," select "ratio > hypothesized ratio" for an upper tail test.

Step 8: Click OK.

Minitab Output

Test and CI for Two Variances

Method

Null hypothesis	$\sigma(\text{First}) / \sigma(\text{Second}) = 1$
Alternative hypothesis	$\sigma(\text{First}) / \sigma(\text{Second}) > 1$
Significance Level	$\alpha = 0.05$

F method was used. This method is accurate for normal data only.

Statistics

Sample	N	StDev	Variance	95% Lower Bound for StDevs
First	12	0.094	0.009	0.070
Second	18	0.054	0.003	0.042

Ratio of standard deviations = 1.741
Ratio of variances = 3.030

95% Confidence Intervals

Method	Lower Bound for StDev Ratio	Lower Bound for Variance Ratio
F	1.121	1.256

Test

Method	DF1	DF2	Test Statistic	P-Value
F	11	17	3.03	0.020

Interpretation of Minitab Output

The Minitab-generated P value is less than the alpha of 0.05, so we reject the null hypothesis in favor of the alternative hypothesis.

One-Tailed F Test (Lower Tail)

Assumptions: Both populations are normally distributed

Null hypothesis (H_0): $H_0: \sigma_1^2 \geq \sigma_2^2$

Alternative hypothesis (H_a): $H_a: \sigma_1^2 < \sigma_2^2$, with rejection criteria $F_0 < F_{1-\alpha,\, n1-1,\, n2-1}$

Test statistic: $F = \dfrac{s_1^2}{s_2^2}$; degrees of freedom $= n_1 - 1, n_2 - 1$

Example

A Six Sigma Black Belt wants to compare the variability in the output of two comparable processes. A sample of 24 is taken from the first process and found to have a standard deviation of 37. A sample of 41 is taken from the second process and found to have a standard deviation of 59. Is the standard deviation of the first population truly lower than that of the second?

Procedure

Step 1: Set the null hypothesis.

$$H_0: \sigma_1^2 \geq \sigma_2^2$$

Step 2: Set the alternative hypothesis.

$$H_a: \sigma_1^2 < \sigma_2^2$$

Step 3: Set the alpha level and determine rejection criteria. The alpha level is 0.05. With degrees of freedom equal to $24 - 1 = 23$ for the numerator and $41 - 1 = 40$ for the denominator, we look at the F distribution table and determine that the critical value is 1.80 and use 1.80 as the rejection value.

Step 4: Determine the test statistic.

$$F = \frac{37^2}{59^2} = 0.393$$

Conclusion

The test statistic of 0.393 is less than the critical value of 1.80, so we reject the null hypothesis. The evidence supports the alternative hypothesis that the variance of the first population is less than the variance of the second population.

Minitab Instructions

Step 1: Go to "Stat > Basic Statistics > σ^2/σ^2 Variances…"

Step 2: Select "Sample variances."

Step 3: Enter the sample size and standard deviation for sample one and sample two.

Step 4: Click "Options."

Step 5: Click "Options" to enter the confidence level. The default setting is 95.0.

Step 6: Under "Ratio," select "(sample 1 standard deviation) / (sample 2 standard deviation)" for standard deviations, or "(Sample variance 1 / Sample variance 2)" if variance is being used.

Step 7: Under "Alternative hypothesis," select "ratio < hypothesized ratio" for a lower tail test.

Step 8: Click OK.

Minitab Output

Test and CI for Two Variances

Method	
Null hypothesis	σ(First) / σ(Second) = 1
Alternative hypothesis	σ(First) / σ(Second) < 1
Significance level	α = 0.05

F method was used. This method is accurate for normal data only.

Statistics

Sample	N	StDev	Variance	95% Upper Bound for StDevs
First	24	37.000	1369.000	49.044
Second	41	59.000	3481.000	72.474

Ratio of standard deviations = 0.627
Ratio of variances = 0.393

95% One-Sided Confidence Intervals

Method	Upper Bound for StDev Ratio	Upper Bound for Variance Ratio
F	0.868	0.753

Test

Method	DF1	DF2	Test Statistic	P-Value
F	23	40	0.39	0.010

Interpretation of Minitab Output

The Minitab-generated *P* value is less than the alpha of 0.05, so we reject the null hypothesis in favor of the alternative hypothesis.

SIMPLE LINEAR REGRESSION

Simple linear regression is used to analyze the relationship between a single continuous independent (predictor) variable and a continuous dependent variable. It is possible to determine a mathematical formula for making predictions based on the regression equation. These variables can be referred to as Y for the predicted variable and X for the variable used to make the prediction. The predicted value (Y) is referred to as the dependent variable, and the variable used as X is called the independent variable. Generally it only makes sense to predict one direction (Clarke and Cooke 1992). For example, we would not want to predict machine speed using surface finish, but we might be interested in the surface finish resulting from the machine speed.

Regression can be used to determine a formula to predict the independent variable's influence on the dependent variable. To use the formula, the variables are multiplied by coefficients. A Pearson correlation coefficient indicates the strength of the linear relationship between the two variables. The formula is

$$r = \frac{\Sigma(x - \bar{x})(y - \bar{y})}{\sqrt{s\,\Sigma(x - \bar{x})^2}\,\sqrt{\Sigma(y - \bar{y})^2}}$$

which is equal to

$$\frac{\Sigma xy - (\Sigma x)(\Sigma y)/n}{\sqrt{\Sigma x^2 - (\Sigma x)^2/n}\,\sqrt{\Sigma y^2 - (\Sigma y)^2/n}}$$

r can range from −1 to 1, with a larger absolute value of *r* indicating a stronger correlation. A negative value indicates there is a negative correlation; the value of the dependent variable decreases as the independent variable increases. A positive number indicates a positive correlation; the value of the dependent variable increases as the independent variable increases. We can also determine where to place a best fit line that passes through the data set. This line is determined by sum of squared errors (or deviations) and it provides us with the prediction equation. The best fit line in Figure 3.9 shows a positive correlation between the two variables. The line in the center is the best fit line. The line should be placed so that the total distance between all data points and the line is minimized. The line shown going from the best fit line to one of the data points represents the difference between the observed value and a predicted value.

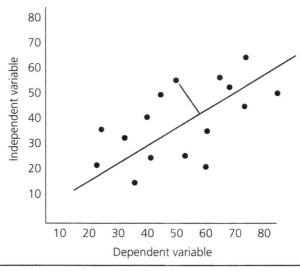

Figure 3.9 Sum of squares line.

The sum of squares for both factors must be determined using the following formulas:

$$\text{Sum of squares X} = SS_X = \Sigma x^2 - \frac{(\Sigma x)^2}{n}$$

$$\text{Sum of squares Y} = SS_Y = \Sigma y^2 - \frac{(\Sigma y)^2}{n}$$

The sum of cross products for the two is equal to:

$$\text{Sum of squares XY} = SS_{XY} = \Sigma xy - \frac{(\Sigma x)(\Sigma y)}{n}$$

With these, the correlation coefficient known as Pearson's r can be determined using the following formula:

$$r = \frac{SS_{XY}}{\sqrt{SS_X}\sqrt{SS_Y}}$$

Predictions can be made using the equation of the best fit line. This is the lowest possible sum of the squares of the distance between data points and the previously mentioned best fit line. The formula is

$$b_0 = \bar{y} - b_1\bar{x}, \text{ with } b_1 = \frac{SCS_{XY}}{SS_X} \text{ and } \bar{x} = \frac{\Sigma x}{n} \text{ and } \bar{y} = \frac{\Sigma y}{n}$$

The equation for the least squares line is $\hat{Y} = b_0 + b_1x$; this is the formula to use to make predictions.

The coefficient of determination is r^2. It indicates how much of the total variability in the dependent variable is explained by the model. For example, an r^2 of 0.44 means that 44% of the observed variation in the dependent variable is explained by the independent variable. The formula is:

$$r^2 = \frac{(SS_{XY})^2}{(SS_X)(SS_Y)}$$

An ANOVA table should be built to facilitate an F-test to determine the significance of the relationship.

$$H_0: \beta = 0$$

$$H_a: \beta \neq 0$$

A coefficient of zero means there is no significant relationship between the variables. A coefficient that is not zero implies a significant relationship.

Example

A process has two factors, factor A and factor B. Factor B is the output, and it is suspected of being influenced by factor A. A process engineer needs to know how to set factor A to get a desired result for factor B. The following table contains the results of 10 trials.

Factor A	Factor B
41	55
31	44
57	70
47	59
45	57
49	59
44	56
54	65
36	47
55	66

Procedure

Step 1: Determine SS_X, SS_Y, and SCP_{XY}. The easiest method is to create a table.

Test	Factor A (X)	Factor B (Y)	XY	X^2	Y^2
1	41	55	2255	1681	3025
2	31	44	1364	961	1936
3	57	70	3990	3249	4900
4	47	59	2773	2209	3481
5	45	57	2565	2025	3249
6	49	59	2891	2401	3481
7	44	56	2464	1936	3136
8	54	65	3510	2916	4225
9	36	47	1692	1296	2209
10	55	66	3630	3025	4356
Summation	459	578	27134	21699	33998

$$SS_x = \Sigma x^2 - \frac{(\Sigma x)^2}{n} = 21{,}669 - \frac{459^2}{10} = 630.9$$

$$SS_Y = \Sigma y^2 - \frac{(\Sigma y)^2}{n} = 33{,}998 - \frac{578^2}{10} = 589.6$$

$$SS_{XY} = \Sigma xy - \frac{(\Sigma x)(\Sigma y)}{n} = 27{,}134 - \frac{(459)(578)}{n} = 603.8$$

Step 2: Determine r, which is

$$r = \frac{SS_{XY}}{\sqrt{SS_x}\sqrt{SS_Y}} = \frac{603.8}{\sqrt{630.9}\sqrt{589.6}} = 0.9899975$$

Step 3: Calculate the least squares line using

$$b_0 = \bar{y} - b_1\bar{x}, \text{ with } b_1 = \frac{SS_{XY}}{SS_x} \text{ and } \bar{x} = \frac{\Sigma x}{n} \text{ and } \bar{y} = \frac{\Sigma y}{n}$$

$$b_1 = \frac{603.8}{630.9} = 0.957$$

$$\bar{y} = \frac{578}{10} = 57.8$$

$$\bar{x} = \frac{459}{10} = 45.9$$

$$b_0 = 57.8 - (0.957)(45.9) = 13.87$$

Step 4: $\hat{Y} = b_0 + b_1 x = 13.87 - (0.957)(x)$. This means we can insert the variable x and calculate the y variable. $\hat{Y} = 13.87 + (0.957)(46) = 57.892$.

Step 5: Calculate the r^2.

$$r^2 = \frac{(SS_{XY})^2}{(SS_x)(SS_Y)} = \frac{(603.8)^2}{(630.9)(589.6)} = 0.98$$

Step 6: Evaluate the results using a hypothesis test to determine whether a linear relationship does exist. The coefficient is represented by a B or β.

$$H_0 : \beta = 0$$

$$H_a : \beta \neq 0$$

$$t = \frac{r}{\sqrt{\dfrac{1 - r^2}{n - 2}}}$$

with degrees of freedom $= n - 2$ and rejection criteria $t < t_{\alpha/2, n-1}$ or $t > t_{\alpha/2, n-1}$. We determine the test statistic is

$$t = \frac{0.989998}{\sqrt{\dfrac{1 - 0.98}{8}}} = 19.8$$

and this is outside the rejection region, so we reject the null hypothesis.

An alternative is to use an *F*-test where *F* is the ratio of the mean squares estimate for the model divided by the mean squares estimate for error.

Conclusion

Ninety-eight percent of the variability in Factor B is the result of Factor A.

Minitab Instructions

Step 1: Go to "Stat > Regression > Fit regression model."

Step 2: Click in the "Responses" box and then select the column for the response.

Step 3: Select "Continuous predictors."

Step 4: Click OK.

Minitab Output

Regression Analysis: C1 versus C2
Analysis of Variance

Source	DF	Adj SS	Adj MS	F-Value	P-Value
Regression	1	618.342	618.342	393.91	0.000
C2	1	618.342	618.342	393.91	0.000
Error	8	12.558	1.570		
Lack-of-Fit	7	10.558	1.508	0.75	0.713
Pure Error	1	2.000	2.000		
Total	9	630.900			

Model Summary

S	R-sq	R-sq(adj)	R-sq(pred)
1.25290	98.01%	97.76%	96.77%

Coefficients

Term	Coef	SE Coef	T-Value	P-Value	VIF
Constant	−13.29	3.01	−4.42	0.002	
C2	1.0241	0.0516	19.85	0.000	1.00

Regression Equation
C1 = −13.29 + 1.0241 C2

Interpretation of Minitab Output

The P value is 0.000, so the null hypothesis is rejected in favor of the alternative hypothesis that there is a significant relationship between A and B. The exact nature of that relationship is that, on average, every one-unit increase in B corresponds with an increase in A of 1.0241 units. Minitab has provided the regression equation: $C1 = -13.929 + 1.0241 \; C2$.

TEST OF DIFFERENCES BETWEEN MEANS (ANOVA)

Analysis of variances (ANOVA) is used to determine if there is a difference in the mean value associated with different levels of a factor (or categorical) variable (Meek et al. 1987). ANOVA is analogous to an independent sample t test, but the t test is limited to having only two levels of the categorical independent variables.

Assumptions: Populations are normally distributed, samples are independent of each other, variances must be the same

Null hypothesis (H_0): $\mu_1 = \mu_2 = \mu_3$

Alternative hypothesis (H_a): Not all μ's are equal, with rejection criteria $F > F_{.05, \, k-1, \, n-k}$

Test statistic: $F = \dfrac{MS_{factor}}{MS_{error}}$

Sum of squares factor = SS (Factor) = $\left[\dfrac{T_1^2}{n_1} + \dfrac{T_2^2}{n_2} + \dfrac{T_3^2}{n_3}\right] - \dfrac{T^2}{n}$,

with $T = T_1 + T_2 + T_3$ and $n = n_1 + n_2 + n_3$

Sum of squares total = SS (Total) = $\Sigma x^2 - \dfrac{T^2}{n}$,

with Σx^2 = the sum of the squared observations

Sum of squares error = SS (Error) = SS (Total) $-$ SS (Factor)

Mean square factor = MS (Factor) = $\dfrac{SS \text{ (Factor)}}{k-1}$

Mean square error = MS (Error) = $\dfrac{SS \text{ (Error)}}{n-k} = S_p^2$

The sum of squares (SS) is used to measure the variation between samples. The letter k is used to represent the number of populations that are being compared. The number of replicates (repeated tests or experiments) in the first population is denoted by n_1, the second population is n_2, and so on; n is the total number of replicates. T_1 is the sum of all observations in the first population, T_2 is the second population, and so on.

The effect of the factor is what is being investigated; it can also be called a treatment, so SS (Factor) can be called SS (Treatment), or SST.

The calculations are summarized in the ANOVA table shown in Table 3.2. The resulting F statistic is evaluated against the F-critical value determined using the F distribution.

Table 3.2 ANOVA table.

Source	d.f.	SS	MS	F
Factor	$k - 1$	SS (Factor)	$\dfrac{SS\ (Factor)}{k - 1}$	$\dfrac{MS\ (Factor)}{MS\ (Error)}$
Error	$n - k$	SS (Error)	$\dfrac{SS\ (Error)}{n - k}$	
Total	$n - 1$	**SS (Total)**		

Example

A Six Sigma Master Black Belt performs an experiment using three levels of a factor. Each experiment is replicated five times.

	Level 1: Catalyst A	Level 2: Catalyst B	Level 3: Catalyst C
	4.7	6.1	5.8
	3.2	4.6	8.3
	3.6	5.6	6.9
	2.4	7.0	4.7
	1.8	4.3	7.1
Means	3.14	5.52	6.56

The following box plot shows the responses for each catalyst. The box plot suggests that the effect will be found to be significant; however, we need to test to be sure.

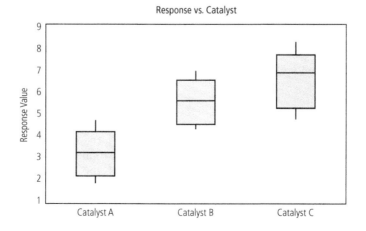

Response vs. Catalyst

Procedure

Step 1: Create an ANOVA table and perform the calculations.

Source	d.f.	SS	MS	F
Factor				
Error				
Total				

Step 2:

$$SS \text{ (Factor)} = \left[\frac{T_1^2}{n_1} + \frac{T_2^2}{n_2} + \frac{T_3^2}{n_3} \right] - \frac{T^2}{n} = \left[\frac{15.7^2}{5} + \frac{27.6^2}{5} + \frac{32.8^2}{5} \right] - \frac{76.1^2}{15} = 30.75$$

$$SS \text{ (Total)} = \Sigma x^2 - \frac{T^2}{n} = 434.15 - \frac{76.1^2}{15} = 48.07$$

$$SS \text{ (Error)} = SS \text{ (Total)} - SS \text{ (Factor)} = 48.07 - 30.75 = 17.32$$

$$MS \text{ (Factor)} = \frac{SS \text{ (Factor)}}{k - 1} = \frac{30.75}{3 - 1} = 15.38$$

$$MS \text{ (Error)} = \frac{SS \text{ (Error)}}{n - k} = S_p^2 = \frac{17.32}{15 - 3} = 1.44$$

Source	d.f.	SS	MS	F
Factor	2	30.75	15.38	10.68
Error	12	17.32	1.44	
Total	**14**	**48.07**		

Conclusion

Use the F table with a 95% confidence level, 2 as a numerator, and 12 as a denominator, written as $F_{.05, 2, 12}$. The F table gives a score of F-critical = 3.89. F-actual = 10.68, which is greater than 3.89, so we reject the null hypothesis. We can conclude that the averages are not the same; however, we do not know which ones are different.

Minitab Instructions

Step 1: Go to "Stat > ANOVA > One-Way…"

Step 2: Select "Response data are in separate column for each factor level" and then select the desired columns.

Step 3: Select options and check the "Assume equal variances" box. The default confidence level is 95.0.

Step 4: Click OK.

Minitab Output

One-way ANOVA: C1; C2; C3

Method

Null hypothesis	All means are equal
Alternative hypothesis	At least one mean is different
Significance level	$\alpha = 0.05$

Equal variances were assumed for the analysis.

Factor Information

Factor	Levels	Values
Factor	3	C1; C2; C3

Analysis of Variance

Source	DF	Adj SS	Adj MS	F-Value	P-Value
Factor	2	30.74	15.369	10.64	0.002
Error	12	17.33	1.444		
Total	14	48.07			

Model Summary

S	R-sq	R-sq(adj)	R-sq(pred)
1.20180	63.94%	57.93%	43.66%

Means

Factor	N	Mean	StDev	95% CI
C1	5	3.140	1.117	(1.969; 4.311)
C2	5	5.520	1.103	(4.349; 6.691)
C3	5	6.560	1.367	(5.389; 7.731)

Pooled StDev = 1.20180

Interpretation of Minitab Output

The resulting Minitab *P* value is less than 0.05, so the null hypothesis should be rejected in favor of the alternative hypothesis. Looking at the confidence intervals generated by Minitab, we can see that Level 1 is significantly different from Level 2 and Level 3. The confidence interval of Level 1 does not overlap with that of Level 2 or Level 3. There is overlap between the confidence levels for Level 2 and Level 3, so we cannot tell if they differ from each other.

Chapter 4
Improve

The Improve phase is where improvements are implemented. Many non-statistical tools and methods are often applied during the Improve phase of a Six Sigma project. For example, lean improvements are implemented to achieve cycle time reductions; theory of constraints (TOC) may be applied here, and small kaizen improvement projects may also take place (Benbow and Kubiak 2009). Often design of experiments (DOE) is implemented to find the ideal settings for critical-to-quality factors identified during the Analyze phase.

DESIGN OF EXPERIMENTS

There are many types of DOE available, such as Latin squares, Plackett-Burman, Taguchi's orthogonal arrays, Box-Behnken, full factorials, and fractional factorials (see Table 4.1). A fractional factorial design does not use all possible combinations of factors; this is in contrast with full factorial designs, which do use all possible combinations of factors. A factor is a process condition that affects the output being investigated. The output being investigated is called the response variable or response. For example, a heat treating experiment aimed at determining the ideal conditions to achieve hardness could use temperature and cooling time as factors, and the response would be hardness. For injection molding, the factors could be mold pressure, moisture, screw speed, dwell time, gate size, cycle time, and material type, and the response would be shrinkage.

Fractional factorial designs use only two levels. In DOE, the level is the setting used in the experiment. The lower levels are often coded with "a–" and the higher levels are represented by "a+." In the heat treating example, the settings could be 210°C and 550°C for heat and 30 seconds and 120 seconds for time. Examples of injection molding levels include a 2 cm gate and a 4 cm gate as well as material from Supplier A and material from Supplier B.

Ideally, each experimental run will be homogenous—that is, comparable to the others. There are, however, situations where the runs are not homogenous. Suppose an injection molding experiment is planned, but there is not sufficient material available from any one supplier. To reduce variability from such influences, blocking can be used. Blocking uses groups or blocks that are similar to each other. This ensures the results reflect variations in the factors and not what was being blocked, such as different suppliers. Box, Hunter, and Hunter (2005) tells us to "block what you can and randomize what you cannot."

Table 4.1 DOE terminology.

Term	Description
Aliasing (confounding)	When the effects of one factor or combination of factors are mixed with the effects of other factors or combinations of factors.
Block	A homogenous unit; used to account for the effects of variation when the entire test group is not homogenous.
Box-Behnken design	A design that screens a large number of main effects using few runs; confounding may be present, but this type of design is economical.
Experimental run	The conditions used during an experiment; can also be called a treatment.
Factor	Something that affects an output (response).
Fractional factorial design	A design that tests only two levels and does not use every possible combination; used when each factor has more than two levels and there are no or negligible interactions between factors.
Full factorial design	A design where all possible combinations of factors are tested. It will yield the best results, but the number of tests required may be unrealistic.
Latin square design	A type of DOE that uses a matrix with each arrangement of conditions only occurring once.
Level	The setting of the factors. For fractional factorial design, only a high and low level are used.
OFAT (one factor at a time)	Testing every possible factor by changing one factor at a time. This is much less efficient than DOE and yields less accurate results because it does not account for interactions.
Plackett-Burman design	A type of two-level fractional factorial screening design used to investigate main effects.
Randomize	Using a randomized run order to help account for variation.
Replicate	Rerunning an experiment in the exact same order and same conditions. This can be thought of as repeating the experiment.
Response	Also known as the response variable, this is the output of the experiment and what we are testing. In Six Sigma terminology, it is the critical-to-quality factor.
Response surface methodology (RSM)	A type of DOE used to determine the optimal response for a process.
Screening DOE	A type of DOE used to reduce a large number of potential factors to a manageable number that can be explored in detail using other designs.
Treatment	The conditions used during an experiment; can also be called an experimental run.

Randomization is also needed when performing DOE; without randomization, there is a risk that the experimental results will reflect unknown changes in the test system over time. For example, an engine test stand may heat up and affect the results. For an experiment with machined cylinders, the cutting tool may become dull over time, resulting in slightly different diameters as each new part is machined. Here, replicates can also be added to a design. This should be done if there is a lot of variation in the process or the resulting measurements. A replicate is a repeat of the experiment using exactly the same test conditions. Replicating each experiment will help to minimize the effects of a variation.

When performing a two-level fractional factorial, a level of resolution must also be selected. This is the degree to which effects are aliased with other effects and can be referred to as confounding; the effects are mixed together and cannot be estimated separately. This is a result of not testing every possible combination. It is a disadvantage of a fractional factorial design; however, not testing every possible combination can have a significant time and financial advantage compared to a full factorial design. The amount of confounding can be controlled by the level of resolution used in the experiment. Generally there are three levels of resolution that are used: resolution III, IV, and V. In a level III resolution, main effects are confounded with two-factor interactions; in a level IV resolution, main effects are confounded with three-factor interactions; and in a level V resolution, there is no confounding of main effects and two-factor interactions (Pries 2009). The confounding problem can be eliminated by performing a full factorial design; however, this requires more experimental runs.

Statistical programs often provide an ANOVA table as well as a main effects plot and an interaction plot. Significant factors can be identified by looking at the factor's P value in the ANOVA table; those with a P value of less than 0.05 are significant. Significant factors are those that influenced the response as they changed from one setting to another. A main effects plot shows the results of changing from one factor level to another, and an interaction plot shows the interactions between the factors (Shankar 2009). An interaction plot with lines that are not parallel is an indication that interactions exist. Main effects can be calculated without the use of a statistical software program by subtracting the average of the responses at the low level from the average of the responses at the high level (Montgomery and Runger 1999).

Example

A Six Sigma Black Belt candidate is attempting to find the optimal settings for a machine. The factors and levels follow.

Factor	Low setting	High setting
Temperature	120	150
Material	B	A
Pressure	85	135

The Six Sigma Black Belt candidate performs a full factorial with one replicate. The results follow.

Run	Temperature	Material	Pressure	Response
1	150	A	135	95.6363
2	120	A	85	90.4674
3	120	A	135	92.3629
4	150	A	135	96.2343
5	120	B	135	91.8775
6	150	B	85	92.3229
7	120	B	135	92.4995
8	120	A	135	92.0589
9	120	B	85	90.5635
10	150	A	85	91.9290
11	120	A	85	89.9334
12	150	B	135	96.3738
13	150	B	85	92.7689
14	150	B	135	95.8418
15	150	A	85	92.5630
16	120	B	85	90.2315

Procedure

Step 1: Determine main effects for temperature.

Temperature low:

$$\frac{90.4674 + 92.3629 + 91.8775 + 93.4995 + 92.0589 + 90.5635 + 89.9334 + 90.2315}{8} =$$

$$\frac{729.9948}{8} = 91.25$$

Temperature high:

$$\frac{95.6363 + 96.2343 + 92.3229 + 91.9290 + 96.3738 + 92.7689 + 95.8418 + 92.5630}{8} =$$

$$\frac{753.67}{8} = 94.2$$

Main effects for temperature: $94.2 - 91.25 = 2.95$

Step 2: Determine main effects for material.

Material low:

$$\frac{91.8775 + 92.3229 + 92.4995 + 90.5635 + 96.3738 + 92.7689 + 95.8418 + 90.2315}{8} =$$

$$\frac{742.479}{8} = 92.80$$

Material high:

$$\frac{95.6363 + 90.4674 + 92.3629 + 96.2343 + 92.0589 + 91.9290 + 89.9334 + 92.5630}{8} =$$

$$\frac{741.185}{8} = 92.65$$

Main effects for material: 92.65 − 92.80 = −0.15

Step 3: Determine main effects for pressure.

Pressure low:

$$\frac{90.4674 + 92.3229 + 90.5635 + 91.9290 + 89.9334 + 92.7689 + 92.5630 + 90.2315}{8} =$$

$$\frac{730.7796}{8} = 91.35$$

Pressure high:

$$\frac{95.6363 + 92.3629 + 96.2343 + 91.8775 + 92.4995 + 92.0589 + 96.3738 + 95.8418}{8} =$$

$$\frac{757.885}{8} = 94.11$$

Main effects for pressure: 94.11 − 91.35 = 2.76

Conclusion

The response variable goes up as temperature and pressure increase; therefore, the optimal settings are 150 for temperature and 135 for pressure.

Minitab Instructions

Step 1: Go to "Stat > DOE > Factorial > Create factorial design…"

Step 2: For type of design, select "2-level factorial (default generators)."

Step 3: Click on "Display Available Designs…" to view possible resolutions for a given number of factors and runs. Then click on "Designs," select the number of replicates, and click OK.

Step 4: Use the "Number of factors" dropdown list to select the number of factors.

Step 5: Select from the available designs based on the required level of resolution. The number of blocks can also be entered here.

Step 6: Click OK and Minitab will create a design.

Step 7: Perform the DOE and enter the results.

Step 8: Go to "Stat > DOE > Factorial > Analyze Factorial Design..."

Step 9: Select the column containing the response variable and select "Graphs."

Step 10: Select "Four in one," then click OK. Click OK again.

Step 11: Go to "Stat > DOE > Factorial > Factorial Plots..." and click OK.

Minitab Output

Factorial Regression: Response versus Temperature, Material, Pressure
Analysis of Variance

Source	DF	Adj SS	Adj MS	F-Value	P-Value
Model	7	68.7075	9.8154	74.21	0.000
Linear	3	65.6780	21.8927	165.53	0.000
Temperature	1	35.0328	35.0328	264.88	0.000
Material	1	0.1047	0.1047	0.79	0.400
Pressure	1	30.5405	30.5405	230.91	0.000
2-Way Interactions	3	3.0273	1.0091	7.63	0.010
Temperature*Material	1	0.0222	0.0222	0.17	0.693
Temperature*Pressure	1	2.9751	2.9751	22.49	0.001
Material*Pressure	1	0.0301	0.0301	0.23	0.646
3-Way Interactions	1	0.0021	0.0021	0.02	0.902
Temperature*Material*Pressure	1	0.0021	0.0021	0.02	0.902
Error	8	1.0581	0.1323		
Total	15	69.7656			

Model Summary

S	R-sq	R-sq(adj)	R-sq(pred)
0.363677	98.48%	97.16%	93.93%

Coded Coefficients

Term	Effect	Coef	SE Coef	T-Value	P-Value	VIF
Constant	92.7290	0.0909	1019.90	0.000	0.000	
Temperature	2.9594	1.4797	0.0909	16.28	0.000	1.00
Material	−0.1618	−0.0809	0.0909	−0.89	0.400	1.00
Pressure	2.7632	1.3816	0.0909	15.20	0.000	1.00
Temperature*Material	−0.0744	−0.0372	0.0909	−0.41	0.693	1.00
Temperature*Pressure	0.8624	0.4312	0.0909	4.74	0.001	1.00
Material*Pressure	0.0867	0.0434	0.0909	0.48	0.646	1.00
Temperature*Material*Pressure	−0.0230	−0.0115	0.0909	−0.13	0.902	1.00

Regression Equation in Uncoded Units

Response = 90.41 − 0.0278 Temperature − 0.39 Material − 0.1000 Pressure + 0.0009 Temperature*Material + 0.001150 Temperature*Pressure + 0.0059 Material*Pressure − 0.000031 Temperature*Material*Pressure

Alias Structure

Factor	Name
A	Temperature
B	Material
C	Pressure

Aliases

I

A

B

C

AB

AC

BC

ABC

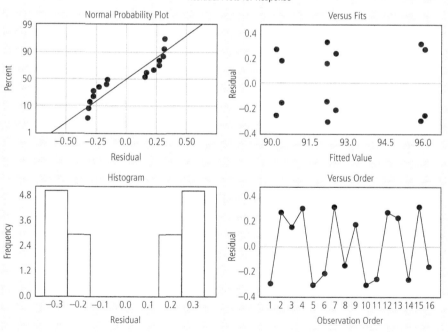

Residual Plots for Response

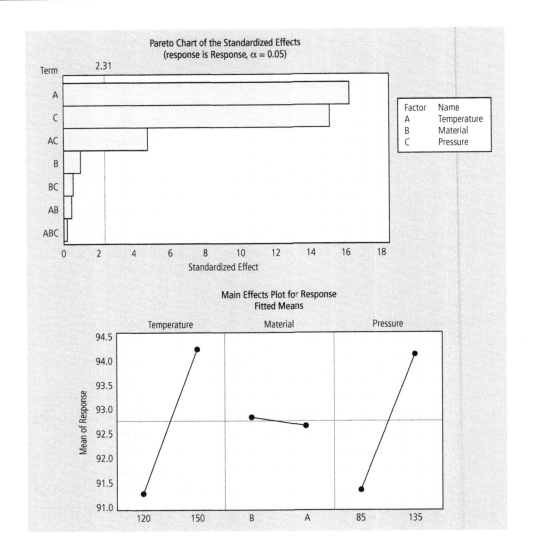

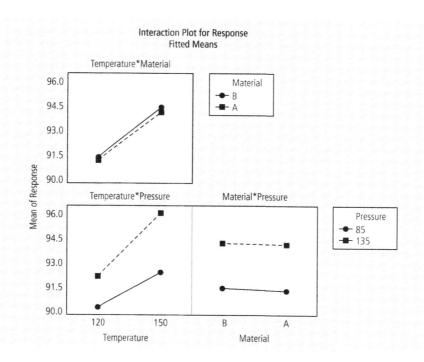

Interpretation of Minitab Output

Minitab provides a Pareto chart of the effects; from this we can tell that only temperature, pressure, and temperature and pressure together are statistically significant. The main effects plot generated by Minitab shows little effect from varying the material. The interaction plot shows an interaction between temperature and pressure.

Chapter 5
Control

The final phase of a Six Sigma project is the Control phase, where actions are taken to ensure that the improvements last and knowledge gained is not lost. Lessons learned should be maintained so that the information garnered over the course of the project is not lost. Control plans and FMEAs should be updated, or created if they do not currently exist. SPC is also implemented for process monitoring.

STATISTICAL PROCESS CONTROL

SPC is a form of process monitoring using statistically determined control limits. Every process has some degree of variation. SPC can be used to identify the two types of process variation, common cause and special cause. Common cause is the normal variation present in the process; special cause is due to an outside influence. A process exhibiting only common cause variation is predictable (Gryna 2001). Corrective action may be needed if special cause variation is present.

Many types of SPC charts are available, depending on the type of data that will be used. Variable data such as measurements like length and diameter can use individual and moving range (I-mR) charts, $\bar{x}$ and S charts, and $\bar{x}$ and R charts. Attribute data, such as pass/fail decisions or percent defective, use c charts, u charts, p charts, and np charts. A flowchart for identifying the correct chart for a particular use is presented in Figure 5.1.

An SPC chart can provide an early warning so that actions can be taken before out-of-specification parts are produced. There are rules that can be applied to the interpretation of SPC data. Following these rules can help prevent tampering and help ensure actions are taken only when common cause variation is present. Figure 5.2 provides the Nelson Rules for interpreting SPC data.

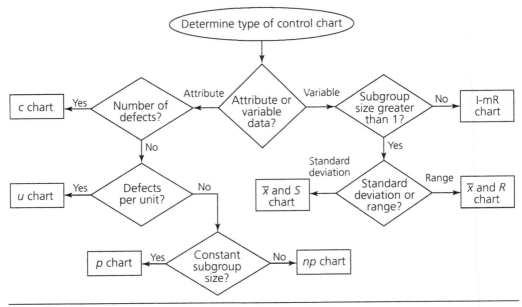

Figure 5.1 Flowchart for determining SPC chart type.

Control Charts for Variable Data

Control charts for variable data use continuous data such as length, weight, and time. The data must be plotted in the order in which they were collected, and a large number of subgroups should be used when setting the control limits. The data need to be free of autocorrelation. This means the value of one result should be independent of the value for the next result.

Rational subgroups are required for control charts for variable data; the variation within a rational subgroup is due to common cause variation. The subgroups must also be homogenous (Besterfield 1998). Material produced on different production lines should not be mixed together when forming a subgroup.

I-mR Chart

An I-mR chart displays individual values and changes in the range from one value to the next. This type of chart is useful when there is insufficient data to form a subgroup, such as when small production batches are used or expensive material is subjected to destructive testing.

Type of data: Continuous values

Subgroup size: 1

Centerline: $\bar{x} = \dfrac{\Sigma x_i}{n}$ and $\overline{mR} = \dfrac{\Sigma MR_i}{n-1}$, with $mR_i = |x_i - x_{i-1}|$ and i equal to

each additional value

Test 1. One point beyond Zone A

Test 2. Nine points in a row in Zone C or beyond

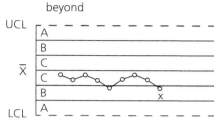

Test 3. Six points in a row steadily increasing or decreasing

Test 4. Fourteen points in a row alternating up and down

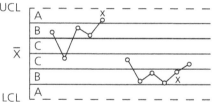

Test 5. Two out of three points in a row in Zone A or beyond

Test 6. Four out of five points in a row in Zone B or beyond

Test 7. Fifteen points in a row in Zone C (above and below centerline)

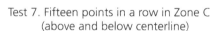

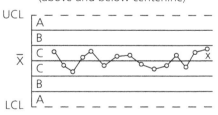

Test 8. Eight points in a row on both sides of centerline with none in Zones C

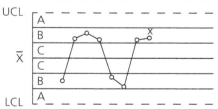

Figure 5.2 Test for special causes.

Source: Lloyd S. Nelson, "Technical Aids," *Journal of Quality Technology* 16, no. 4 (October 1984): 238–239.

Abbreviations: $\bar{x}$ = process average, mR = median of moving range

Upper control limit for x: $\bar{x} + E_2 m\bar{R}$

Upper control limit for mR: $D_4 m\bar{R}$

Lower control limit for x: $\bar{x} - E_2 m\bar{R}$

Lower control limit for mR: $D_3 m\bar{R}$

Example

A Six Sigma Green Belt samples 11 different 50 kilogram bags of molding machine plastic granules. The weights of the bags are 50.12, 49.64, 50.11, 49.87, 49.99, 49.68, 50.38, 50.38, 49.40, 40.70, and 49.77.

Procedure

Step 1: Find the centerline for x by finding the average of the data set.

$$(50.12 + 49.64 + 50.11 + 49.87 + 49.99 + 49.68 + 50.38 + 50.38 + 49.40 + 49.70 + 49.77) / 11 = 49.9.$$

Step 2: To identify the upper control limit for x, find E_2 for subgroup size 2 in the SPC constants table (see Appendix D). E_2 is equal to 2.660.

Step 3: Find the centerline for $m\bar{R}$, which is equal to $\dfrac{\Sigma MR_i}{n-1}$, by creating a table. Starting with sample 2, subtract the previous value from the current value.

Sample	Weight	Moving range
1	50.12	
2	49.64	0.48
3	50.11	0.47
4	49.87	0.24
5	49.99	0.12
6	49.68	0.31
7	50.38	0.70
8	50.38	0.00
9	49.40	0.98
10	49.70	0.30
11	49.77	0.07
Total		3.67

Step 4: Divide the total for moving range by $n - 11$: 3.67 / (11 − 1) = 0.367.

Step 5: Calculate the upper control limit for x ($\bar{x} + E_2 m\bar{R}$): 49.9 + 2.660(0.367) = 50.89.

Step 6: Calculate the lower control limit for x ($\bar{x} - E_2 m\bar{R}$): $49.9 - 2.660(0.367) = 48.92$.

Step 7: Find the upper control limit for mR by looking up D_4 for subgroup size 2 and then calculating $D_4 m\bar{R}$: $3.267(0.367) = 1.19$.

Step 8: Find the lower control limit for mR by looking up D_3 for subgroup size 2 and then calculating $D_3 m\bar{R}$: $0.0(0.367) = 0$.

Step 9: Draw the individual and moving range charts with the centerlines and upper and lower control limits and then plot the data.

Minitab Instructions

Step 1: Go to "Stat > Control Charts > Variables Charts for Individuals > I-MR…"

Step 2: Select the column containing the data for "Variables" and click OK.

Minitab Output

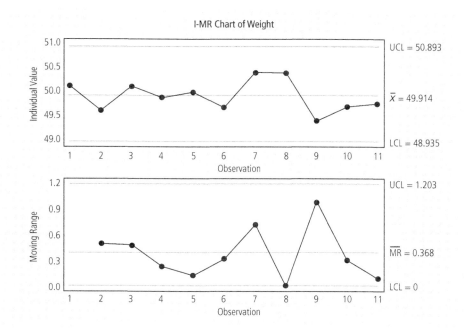

Interpretation of Minitab Output

The I-mR chart shows only common cause variation.

$\bar{x}$ *and* S *Chart*

The $\bar{x}$ and S chart uses the subgroup average and standard deviation of the sample. Using an $\bar{x}$ and S chart requires a subgroup size greater than 1 and is ideal for large subgroups.

Type of data: Continuous values

Subgroup size: >1

Centerline: $\bar{\bar{x}} = \frac{\Sigma \bar{x}}{k}$ and $\bar{S} = \frac{\Sigma \bar{s}}{k}$

Abbreviations: $\bar{x}$ = process average, S = process standard deviation, k = number of subgroups

Upper control limit: $\bar{x} = \bar{\bar{x}} + A_3 \bar{s}$ and $\bar{s} = B_4 \bar{s}$

Lower control limit: $\bar{x} = \bar{\bar{x}} - A_3 \bar{s}$ and $\bar{s} = B_3 \bar{s}$

Example

A process engineer collects 5 subgroups of 5 parts from a milling operation. The data are presented in the following table.

Subgroup 1	Subgroup 2	Subgroup 3	Subgroup 4	Subgroup 5
115.113	113.391	116.218	114.952	114.489
113.562	115.638	115.040	114.617	113.634
115.641	115.746	117.290	113.637	114.395
115.691	114.053	118.693	113.494	113.888
112.292	115.373	116.082	116.326	112.707

Procedure

Step 1: Determine $\bar{\bar{x}}$ and $\bar{s}$ by creating a table to determine the average and standard deviation of each subgroup.

Subgroup 1	Subgroup 2	Subgroup 3	Subgroup 4	Subgroup 5	Average	Standard deviation
115.113	113.391	116.218	114.952	114.489	114.06	1.36
113.562	115.638	115.040	114.617	113.634	114.84	1.06
115.641	115.746	117.290	113.637	114.395	116.86	1.39
115.691	114.053	118.693	113.494	113.888	114.61	1.15
112.292	115.373	116.082	116.326	112.707	113.82	0.717
				Average	114.84	1.135

The average of the averages is the centerline $\bar{\bar{x}}$, 114.84. The average of the standard deviations is the centerline $\bar{s}$, 1.135.

Step 2: Find the upper control limit for $\bar{x}$, which is equal to $\bar{\bar{x}} + A_3 \bar{s}$. Find A_3 for a subgroup of 5 in the SPC constant chart. 114.84 + 1.342(1.135) = 116.36.

Step 3: Find the lower control limit for $\bar{x}$, which is equal to $\bar{\bar{x}} - A_3 \bar{s}$. Find A_3 for a subgroup of 5 in the SPC constant chart. 114.84 − 1.342(1.135) = 113.32.

Step 4: Find the upper control limit for $\bar{s}$, which is equal to $B_4\bar{s}$. Find the constant for B_4 with a subgroup size of 5 in the SPC constant chart. 2.089(1.145) = 2.39.

Step 5: Find the lower control limit for $\bar{s}$, which is equal to $B_3\bar{s}$. Find the constant for B_3 with a subgroup size of 5 in the SPC constant chart. 0.0(1.145) = 0.0.

Step 6: Draw the $\bar{x}$ and S charts with the centerlines and upper and lower control limits and then plot the data.

Minitab Instructions

Step 1: Go to "Stat > Control Charts > Variables Charts for Subgroups > Xbar-S..."

Step 2: Select the column containing the data.

Step 3: Enter the subgroup size and click OK.

Minitab Output

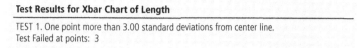

Test Results for Xbar Chart of Length

TEST 1. One point more than 3.00 standard deviations from center line.
Test Failed at points: 3

WARNING If graph is updated with new data, the results above may no longer be correct.

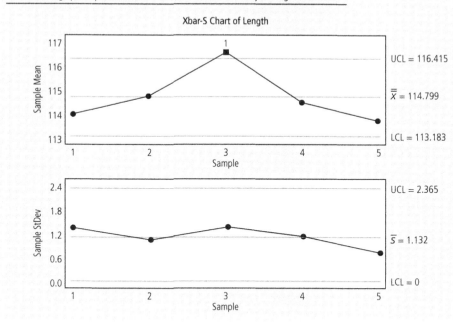

Interpretation of Minitab Output

The $\bar{x}$ chart shows the presence of special cause variation. There is no special cause variation in the S chart.

$\bar{x}$ *and* **R** *Chart*

An $\bar{x}$ and R chart is used to monitor the variability between subgroups of constant size. The subgroup size must be less than 10; typically a sample size of 5 is used.

Type of data: Continuous values

Subgroup size: <10 (typically 5)

Centerline: $\overline{\overline{X}} = \dfrac{\Sigma \overline{x}}{k}$ and $\overline{R} = \dfrac{\Sigma \overline{R}}{k}$

Abbreviations: $\overline{x}$ = process average, R = range

Upper control limit: $\overline{x} = \overline{\overline{x}} + A_2\overline{R}$ and $\overline{R} = D_4\overline{R}$

Lower control limit: $\overline{x} = \overline{\overline{x}} - A_2\overline{R}$ and $\overline{R} = D_3\overline{R}$

Example

A Six Sigma Black Belt collects 5 subgroups of 5 on the length of stamped components. The data are displayed in the following table.

Subgroup 1	Subgroup 2	Subgroup 3	Subgroup 4	Subgroup 5
23.537	23.326	23.352	23.576	23.488
23.456	23.673	23.150	23.2062	23.392
23.302	23.460	23.687	23.771	23.572
23.819	23.346	23.778	23.128	23.519
23.484	23.684	23.410	23.307	23.068

Procedure

Step 1: Determine $\overline{x}$ and R by creating a table to determine the average and range of each subgroup.

Subgroup 1	Subgroup 2	Subgroup 3	Subgroup 4	Subgroup 5	Average	Range
23.537	23.326	23.352	23.576	23.488	23.52	0.50
23.456	23.673	23.150	23.2062	23.392	23.50	0.36
23.302	23.460	23.687	23.771	23.572	23.38	0.54
23.819	23.346	23.778	23.128	23.519	23.40	0.64
23.484	23.684	23.410	23.307	23.068	23.41	0.50
				Average	23.44	0.51

The average of the averages is the centerline $\overline{\overline{x}}$, 23.44. The average of the ranges is the centerline $\overline{R}$, 0.51.

Step 2: Find the upper control limit for $\overline{x}$, which is equal to $\overline{\overline{x}} + A_2\overline{R}$. Find A_2 for a subgroup of 5 in the SPC constant chart. 23.44 + 0.577(0.51) = 23.74.

Step 3: Find the lower control limit for $\bar{x}$, which is equal to $\bar{\bar{x}} - A_2\bar{R}$. Find A_2 for a subgroup of 5 in the SPC constant chart. $23.44 - 0.577(0.51) = 23.15$.

Step 4: Find the upper control limit for $\bar{R}$, which is equal to $D_4\bar{R}$. Find the constant for D_4 with a subgroup size of 5 in the SPC constant chart. $2.004(0.51) = 1.02$.

Step 5: Find the lower control limit for $\bar{R}$, which is equal to $D_3\bar{R}$. Find the constant for D_3 with a subgroup size of 5 in the SPC constant chart. $0.0(0.51) = 0.0$.

Step 6: Draw the $\bar{x}$ and R charts with the centerlines and upper and lower control limits and then plot the data.

Minitab Instructions

Step 1: Go to "Stat > Control Charts > Variables Charts for Subgroups > Xbar-R…"

Step 2: Select the column containing the data.

Step 3: Enter the subgroup size and click OK.

Minitab Output

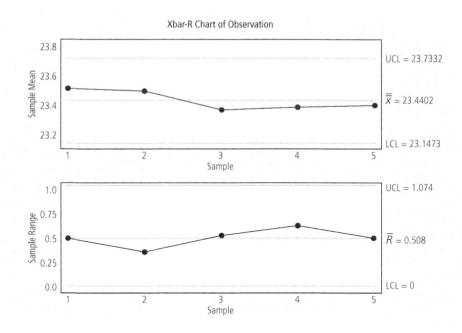

Interpretation of Minitab Output

The average and range charts show only common cause variation.

Control Charts for Attribute Data

Control charts for attribute data are used for count items, such as the number of defects or the number of defective units. When using control charts for attribute data it is important to use a sample size that is large enough to include defective units or defects. A small subgroup size is not appropriate if the number of defects or defective units is very low. For example, a subgroup size of 10 has little chance of including a defect if the process has a failure rate of 1%.

The *p* chart and *np* chart are used for binomial data, such as when items are classified as passed or failed. The *c* chart and *u* chart are used for count data that follow a Poisson distribution (George et al. 2005).

c *Chart*

A *c* chart is used for the number of defects per item when the subgroup size is constant. An appliance with a scratch, a dent, and a missing switch would count as three defects.

Type of data: Number of defects (attribute)

Subgroup size: Constant size

Centerline: $\bar{c}$

Abbreviation: c = number of defects

Upper control limit: $\bar{c} + 3\sqrt{\bar{c}}$

Lower control limit: $\bar{c} - 3\sqrt{\bar{c}}$

Example

A quality technician on a Six Sigma team collects 9 subgroups of 95 each from a process. The subgroups contain the following number of defects: 16, 24, 14, 10, 16, 12, 11, 15, and 9.

Procedure

Step 1: Determine the centerline $\bar{c}$ by averaging the number of defects.

$$\bar{c} = (16 + 24 + 14 + 10 + 16 + 12 + 11 + 15 + 9) / 9 = 14.11$$

Step 2: Find the upper control limit, which is $\bar{c} + 3\sqrt{\bar{c}} = 14.11 + 3\sqrt{14.11} = 25.38$.

Step 3: Find the lower control limit, which is $\bar{c} - 3\sqrt{\bar{c}} = 14.11 - 3\sqrt{14.11} = 2.84$.

Step 4: Draw the $\bar{c}$ chart with the centerline and upper and lower control limits and then plot the data.

Minitab Instructions

Step 1: Go to "Stat > Control Charts > Attributes Charts > C..."

Step 2: Select the column containing the data for "Variables" and click OK.

Minitab Output

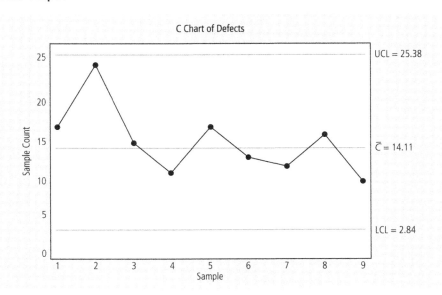

C Chart of Defects

Interpretation of Minitab Output

The number of defects could vary greatly, but the process is in control.

u *Chart*

The u chart is used for the number of defects per unit; for example, a metal sheet with three dents would count as three defects, and a work order with one line of information missing and one line filled out incorrectly would count as two defects. The subgroup size for a u chart can vary.

Type of data: Number of defects per unit (attribute)

Subgroup size: Size can vary

Centerline: $\bar{u}$

Abbreviation: $\bar{u}$ = total number of defects found in all subgroups / total number of items checked in all subgroups

Upper control limit: $\bar{u} + 3\sqrt{\dfrac{\bar{u}}{n}}$

Lower control limit: $\bar{u} - 3\sqrt{\dfrac{\bar{u}}{n}}$

Example

A Six Sigma Black Belt takes 8 subgroups from a process. The subgroup sizes and number of defects in each subgroup is shown in the following table.

Sample	Defects
8	4
11	5
8	3
8	4
11	6
9	6
11	1
10	2

Procedure

Step 1: Determine centerline $\bar{u}$ by dividing the total number of defects by the total number of items checked. $\bar{u} = (4 + 5 + 3 + 4 + 6 + 6 + 1 + 2) / (8 + 11 + 8 + 8 + 11 + 9 + 11 + 10) = 0.408$.

Step 2: Find the upper control limit for each subgroup, which is equal to $\bar{u} + 3\sqrt{\dfrac{\bar{u}}{n}}$, with n = the size of the subgroup.

$$\text{Subgroup 1} = 0.408 + 3\sqrt{\frac{0.408}{8}} = 1.09$$

$$\text{Subgroup 2} = 0.408 + 3\sqrt{\frac{0.408}{11}} = 0.99$$

$$\text{Subgroup 3} = 0.408 + 3\sqrt{\frac{0.408}{8}} = 1.09$$

$$\text{Subgroup 4} = 0.408 + 3\sqrt{\frac{0.408}{8}} = 1.09$$

$$\text{Subgroup 5} = 0.408 + 3\sqrt{\frac{0.408}{11}} = 0.99$$

$$\text{Subgroup 6} = 0.408 + 3\sqrt{\frac{0.408}{9}} = 1.04$$

$$\text{Subgroup 7} = 0.408 + 3\sqrt{\frac{0.408}{11}} = 0.99$$

$$\text{Subgroup } 8 = 0.408 + 3\sqrt{\frac{0.408}{10}} = 1.01$$

Step 3: Find the lower control limit for each subgroup, which is equal to $\bar{u} - 3\sqrt{\frac{\bar{u}}{n}}$, with n = the size of the subgroup.

$$\text{Subgroup } 1 = 0.408 - 3\sqrt{\frac{0.408}{8}} = -0.27$$

The lower control limits are all negative numbers and a negative control limit is not possible, so use 0 as the lower control limit.

Step 4: Draw the $\bar{u}$ chart with the centerline and upper and lower control limits and then plot the data.

Minitab Instructions

Step 1: Go to "Stat > Control Charts > Attributes Charts > U…"

Step 2: Select the column containing the data for "Variables."

Step 3: Enter the column containing the subgroup sizes and click OK.

Minitab Output

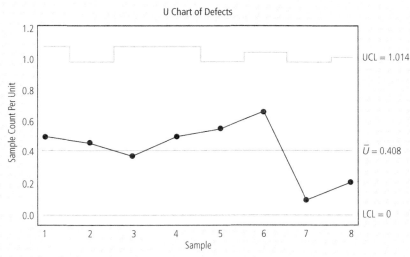

Tests performed with unequal sample sizes.

Interpretation of Minitab Output

The upper control limit varies due to changes in sample size; however, the process is in control.

p *Chart*

A *p* chart is used for the proportion rejected. Subgroups can be of various sizes. For example, the first subgroup may have 86 items with 13 rejected for a proportion of 0.12, and the second subgroup may have 72 items with 8 rejected for a proportion of 0.11.

Type of data: Fraction of units defective (attribute)

Subgroup size: Can vary, but greater than 50

Centerline: $\bar{p}$

Abbreviation: p = proportion defective

Upper control limit: $\bar{p} + 3\dfrac{\sqrt{\bar{p}(1-\bar{p})}}{\sqrt{\bar{n}}}$

Lower control limit: $\bar{p} - 3\dfrac{\sqrt{\bar{p}(1-\bar{p})}}{\sqrt{\bar{n}}}$

Example

A quality inspector collects 12 subgroups of varying sizes from a process. The results are shown in the following table.

Sampled	Rejected
52	5
58	3
62	6
58	3
54	16
61	4
60	3
53	5
57	3
55	4
54	2
61	6

Procedure

Step 1: Create a table to determine the proportion rejected for each subgroup and then total the proportions rejected.

Sampled	Rejected	Proportion rejected
52	5	0.096
58	3	0.052
62	6	0.097
58	3	0.052
54	16	0.296
61	4	0.066
60	3	0.050
53	5	0.094
57	3	0.053
55	4	0.073
54	2	0.037
61	6	0.098
	Total	1.064

Step 2: Find the centerline $\bar{p}$, which is the average of proportions rejected. The total of proportions rejected divided by the number of subgroups is 1.064 / 12 = 0.088.

Step 3: Find the average of the sample sizes $\bar{n}$, which is the total number of samples taken divided by the number of subgroups.

$$(52 + 58 + 62 + 58 + 54 + 61 + 60 + 53 + 57 + 55 + 54 + 61) / 12 = 57.083$$

Step 4: Find the upper control limit.

$$\bar{p} + 3 \frac{\sqrt{\bar{p}(1 - \bar{p})}}{\sqrt{\bar{n}}} = 0.088 + 3 \frac{\sqrt{0.088(1 - 0.088)}}{\sqrt{57.083}} = 0.200$$

Step 5: Find the lower control limit.

$$\bar{p} - 3 \frac{\sqrt{\bar{p}(1 - \bar{p})}}{\sqrt{\bar{n}}} = 0.088 - 3 \frac{\sqrt{0.088(1 - 0.088)}}{\sqrt{57.083}} = -0.24$$

The lower control limit cannot be a negative number, so use 0 as the lower control limit.

Step 6: Draw the p chart with the centerline and upper and lower control limits and then plot the data.

Minitab Instructions

Step 1: Go to "Stat > Control Charts > Attributes Charts > P..."

Step 2: Select the column containing the data for "Variables."

Step 3: Enter the column containing the subgroup sizes and click OK.

Minitab Output

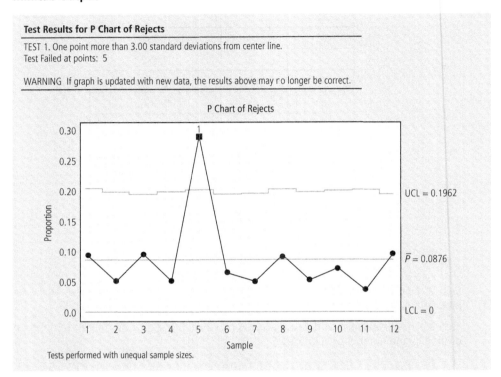

Test Results for P Chart of Rejects

TEST 1. One point more than 3.00 standard deviations from center line.
Test Failed at points: 5

WARNING If graph is updated with new data, the results above may ro longer be correct.

P Chart of Rejects

UCL = 0.1962

$\bar{P}$ = 0.0876

LCL = 0

Tests performed with unequal sample sizes.

Interpretation of Minitab Output

Minitab shows one point failing.

np *Chart*

An *np* chart is used for the number of items rejected in a subgroup. This could be trim panels with scratches or bills with errors. The *np* chart uses items rejected, not the number of failures per item. For example, a bill with three errors would count as just one reject. An *np* chart can only be used when the subgroup sizes are the same. An *np* chart can be useful for comparing the performance of a process before and after improvements have been implemented.

Type of data: Number of defective units (attribute)

Subgroup size: Constant and greater than 50

Centerline: $\overline{np}$

Abbreviations: p = proportion defective, np = number of defective units

Upper control limit: $\overline{np} + 3\sqrt{\overline{np}(1-\overline{p})}$

Lower control limit: $\overline{np} - 3\sqrt{\overline{np}(1-\overline{p})}$

Example

A Six Sigma Black Belt receives a quality report for 10 subgroups of sample size 80 each. The number rejected for each subgroup is listed in the following table.

Inspected	Rejected
80	2
80	5
80	4
80	6
80	5
80	3
80	17
80	17
80	4
80	5

Procedure

Step 1: Create a table to find the total number inspected and the total number rejected.

	Inspected	Rejected
	80	2
	80	5
	80	4
	80	6
	80	5
	80	3
	80	17
	80	17
	80	4
	80	5
Total	800	68

Step 2: Find $\overline{p}$ by dividing the total number rejected by the total number inspected: $\overline{p} = 68 / 800 = 0.085$.

Step 3: Find the centerline $\overline{np}$ by multiplying the subgroup size by the total number of rejects divided by the total number of items inspected.

$$\overline{np} = 80(68 / 800) = 6.8$$

Step 4: Calculate the upper control limit.

$$\overline{np} + 3\sqrt{\overline{np}(1 - \overline{p})} = 6.8 + 3\sqrt{6.8(1 - 0.085)} = 14.28$$

Step 5: Calculate the lower control limit.

$$\overline{np} - 3\sqrt{\overline{np}(1 - \overline{p})} = 6.8 - 3\sqrt{6.8(1 - 0.085)} = -0.683$$

A negative number cannot be used as a lower control limit, so use 0.

Step 6: Draw the $\overline{np}$ chart with the centerline and upper and lower control limits and then plot the data.

Minitab Instructions

Step 1: Go to "Stat > Control Charts > Attributes Charts > NP..."

Step 2: Select the column containing the data for "Variables."

Step 3: Enter the subgroup size and click OK.

Minitab Output

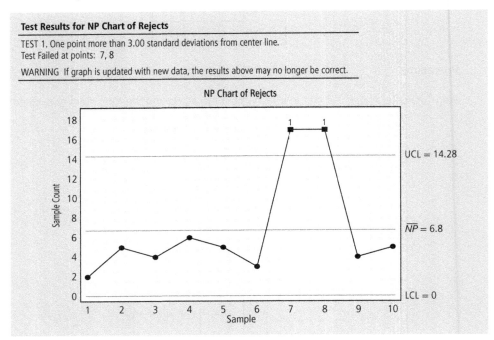

Interpretation of Minitab Output

Minitab shows tests failing at two points.

Chapter 6
Conclusion

Six Sigma Black Belts need to be experts in statistics, perhaps not at the level of a practicing statistician, but still sufficiently knowledgeable to be able to identify and apply the correct statistical method. This book is intended to help Black Belts to apply statistical methods like an expert. The easy to follow examples are intended to help the Black Belt who has not needed a particular method in a long time and as the Black Belt gains proficiency in each of the statistical methods described here, he or she can start to skip the detailed examples and use this book as a quick reference to find formulas and the assumptions that must be met to use the formulas.

Appendix A
Hypothesis Test Selection

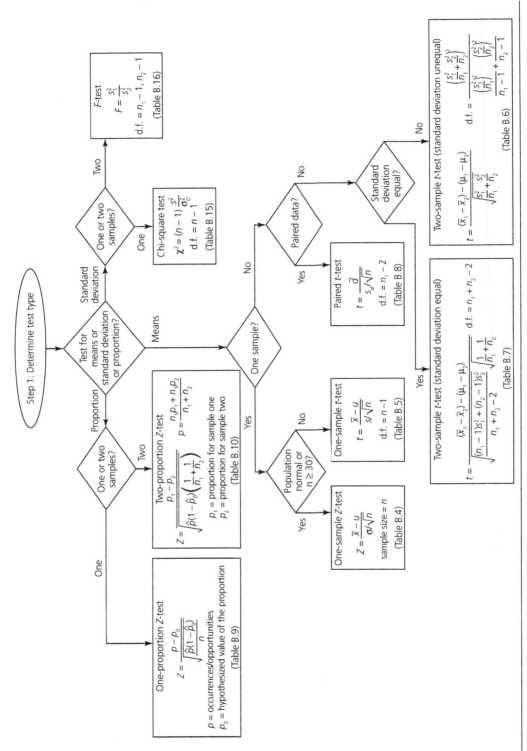

Figure A.1 Decision tree for selecting statistical tests.

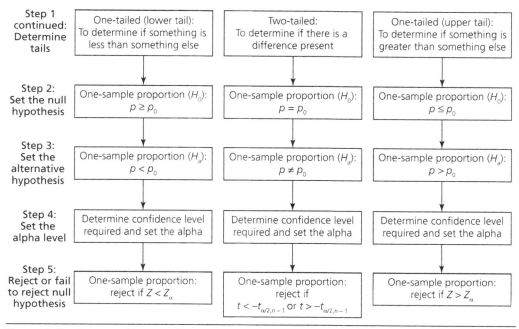

Figure A.2   One-sample proportion Z-test.

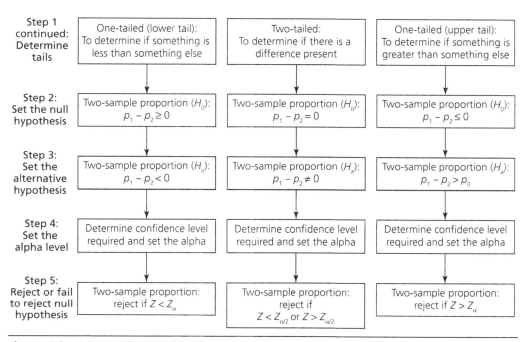

Figure A.3   Two-sample proportion Z-test.

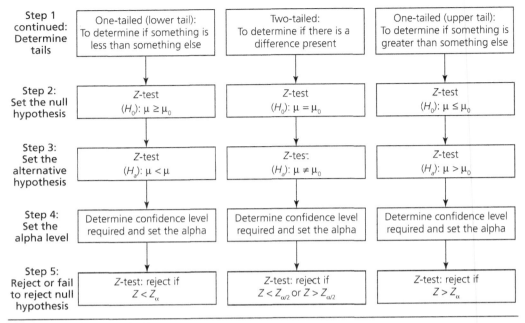

Figure A.4   One-sample Z-test.

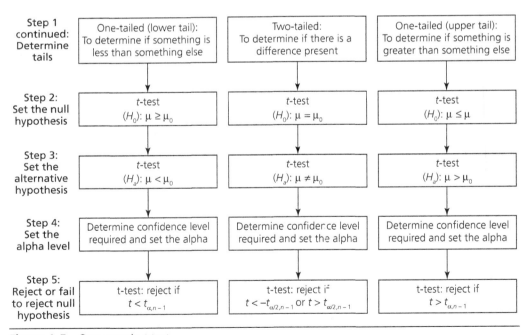

Figure A.5   One-sample t-test.

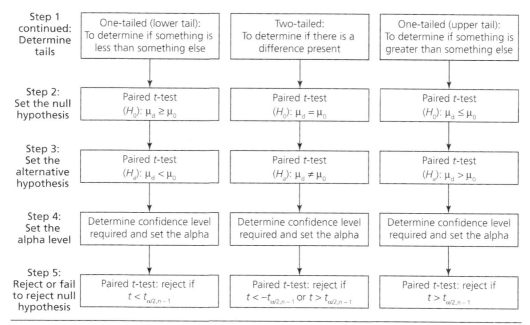

Figure A.6 Paired *t*-test.

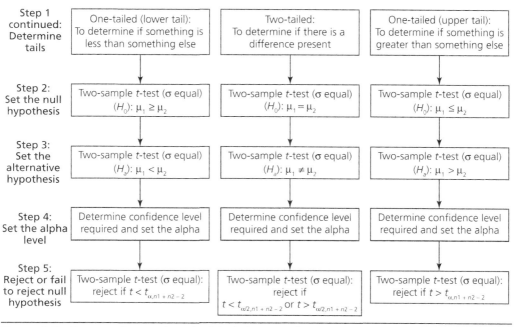

Figure A.7 Two-sample *t*-test (standard deviations equal).

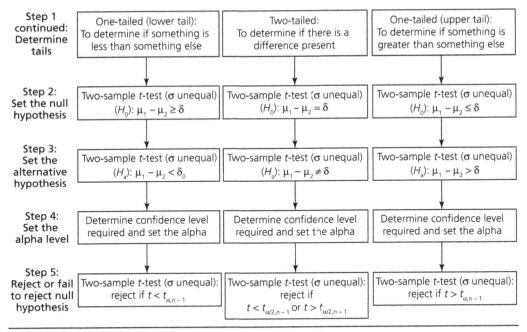

Figure A.8   Two-sample *t*-test (standard deviations unequal).

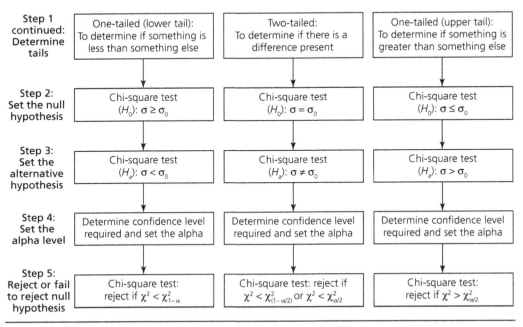

Figure A.9   Chi-square test.

Step 1 continued: Determine tails	One-tailed (lower tail): To determine if something is less than something else	Two-tailed: To determine if there is a difference present	One-tailed (upper tail): To determine if something is greater than something else
Step 2: Set the null hypothesis	F-test (H_0): $\sigma \geq \sigma_0$	F-test (H_0): $\sigma = \sigma_0$	F-test (H_0): $\sigma \leq \sigma_0$
Step 3: Set the alternative hypothesis	F-test (H_a): $\sigma < \sigma_0$	F-test (H_a): $\sigma \neq \sigma_0$	F-test (H_a): $\sigma > \sigma_0$
Step 4: Set the alpha level	Determine confidence level required and set the alpha	Determine confidence level required and set the alpha	Determine confidence level required and set the alpha
Step 5: Reject or fail to reject null hypothesis	F-test: reject if $F_0 < F_{1\alpha,n1-1,n2-2}$	F-test: reject if $F_0 > F_{\alpha/2,n1-1,n2-2}$ or reject if $F_0 < F_{1\alpha/2,n1-1,n2-2}$	F-test: reject if $F_0 > F_{1-\alpha/2,n1-1,n2-2}$

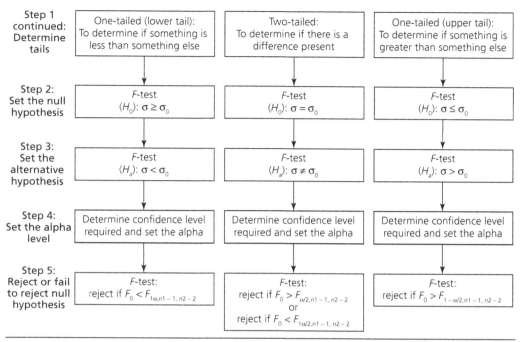

Figure A.10   *F*-test.

Appendix B
Quick Reference for Formulas

Table B.1 Probability.

Addition rule for union of two non-mutually-exclusive events: $P(A \cup B) = P(A) + P(B) - P(A \cap B)$
Addition rule for union of two mutually exclusive events: $P(A \cup B) = P(A) + P(B)$
Multiplication rule for the intersection of two dependent events: $P(A \cap B) = P(A) \times P(B|A)$
Multiplication rule for the intersection of two independent events: $P(A \cap B) = P(A) \times P(B)$
Binomial distribution: $p(k) = \dfrac{n!}{k!(n-k)!} P^k (1 - P)^{n-k}$
Bayes' theorem: $P(A|B) = \dfrac{P(A)P(B|A)}{P(B|A)P(A) + P(B|\sim A)P(\sim A)}$

Table B.2 Descriptive statistics.

Mean:
$$\bar{x} = \frac{\Sigma x_i}{n}$$
Sample variance:
$$S^2 = \frac{1}{1-n} \Sigma (x_i - \bar{x})^2$$
Population variance:
$$S^2 = \frac{\Sigma (x_i - \bar{x})^2}{n-1}$$
Sample standard deviation:
$$S = \sqrt{s^2}$$
Population standard deviation:
$$\sigma = \sqrt{\sigma^2}$$

Table B.3 Skew and kurtosis.

Skew:
$$Sk = \frac{3(\bar{x} - Md)}{s}$$
Kurtosis:
$$\text{Kurtosis} = \frac{n}{(n-1)(n-2)} \Sigma \left(\frac{x - \bar{x}}{s} \right)^3 = \frac{n}{(n-1)(n-2)} \Sigma (x_i - \bar{x}) / s^3$$

Table B.4 *Z*-test.

One-sample hypothesis tests for mean with large sample size, any population, and known σ; or small sample size, normally distributed population, and known σ
Type: Mean
Assumptions: One sample with $n \geq 30$; population type: any; σ: known or One sample with $n < 30$; population type: normal; σ: known
Test statistic: $$Z = \frac{\bar{x} - u}{\sigma/\sqrt{n}}, \text{ sample size} = n$$
Two-tailed hypothesis: Null hypothesis (H_0): $\mu = \mu_0$ Alternative hypothesis (H_a): $\mu \neq \mu_0$ Rejection criteria: $Z < Z_{\alpha/2}$ or $Z > Z_{\alpha/2}$
One-tailed hypothesis (upper/right tail): Null hypothesis (H_0): $\mu \leq \mu_0$ Alternative hypothesis (H_a): $\mu > \mu_0$ Rejection criteria: $Z > Z_{\alpha}$
One-tailed hypothesis (lower/left tail): Null hypothesis (H_0): $\mu \geq \mu_0$ Alternative hypothesis (H_a): $\mu < \mu_0$ Rejection criteria: $Z < Z_{\alpha}$

Table B.5 One-sample *t*-test.

One-sample hypothesis tests for mean with small sample size, normally distributed population, and unknown σ
Type: Mean
Assumptions: Sample size: $n < 30$; population type: normal; σ: unknown
Test statistic: $t = \dfrac{\bar{x} - \mu}{s/\sqrt{n}}$, degrees of freedom $= n - 1$
Two-tailed hypothesis: Null hypothesis (H_0): $\mu = \mu_0$ Alternative hypothesis (H_a): $\mu \neq \mu_0$ Rejection criteria: $t < t_{\alpha/2, n-1}$ or $t > t_{\alpha/2, n-1}$
One-tailed hypothesis (upper/right tail):   Null hypothesis (H_0): $\mu \leq \mu_0$ Alternative hypothesis (H_a): $\mu > \mu_0$ Rejection criteria: $t > t_{\alpha, n-1}$
One-tailed hypothesis (lower/left tail):   Null hypothesis (H_0): $\mu \geq \mu_0$ Alternative hypothesis (H_a): $\mu < \mu_0$ Rejection criteria: $t < t_{\alpha, n-1}$

Table B.6   Two-sample *t*-test (unequal standard deviations).

Two-sample hypothesis tests for mean with small sample size, independent samples, normally distributed population, and unknown and unequal standard deviations
Type: Mean
Assumptions: Sample: n_1. $n_2 < 30$, independent samples; population: normal; standard deviations: unknown and unequal
Test statistic: $$t = \frac{(\bar{x}_1 - \bar{x}_2) - (\mu_1 - \mu_2)}{\sqrt{\dfrac{s_1^2}{n_1} + \dfrac{s_2^2}{n_2}}} \text{ , degrees of freedom} = \frac{\left(\dfrac{s_1^2}{n_1} + \dfrac{s_2^2}{n_2}\right)^2}{\dfrac{\left(\dfrac{s_1^2}{n_1}\right)^2}{n_1 - 1} + \dfrac{\left(\dfrac{s_2^2}{n_2}\right)^2}{n_2 - 1}}$$
Two-tailed hypothesis: Null hypothesis (H_0): $\mu_1 - \mu_2 = \delta$ (δ is the hypothesized difference between μ_1 and μ_2) Alternative hypothesis (H_a): $\mu_1 - \mu_2 \neq \delta$ Rejection criteria: $t < t_{\alpha/2,n-1}$ or $t > t_{\alpha/2,n-1}$
One-tailed hypothesis (upper/right tail): Null hypothesis (H_0): $\mu_1 - \mu_2 \leq \delta$ (δ is the hypothesized difference between μ_1 and μ_2) Alternative hypothesis (H_a): $\mu_1 - \mu_2 > \delta$ Rejection criteria: $t > t_{\alpha,n-1}$
One-tailed hypothesis (lower/left tail): Null hypothesis (H_0): $\mu_1 - \mu_2 \geq \delta$ (δ is the hypothesized difference between μ_1 and μ_2) Alternative hypothesis (H_a): $\mu_1 - \mu_2 < \delta$ Rejection criteria: $t < t_{\alpha,n-1}$

Table B.7   Two-sample *t*-test (equal standard deviations).

Hypothesis tests for mean with small sample size, independent samples, any population shape, and unknown but equal standard deviations
Type: Mean
Assumptions: Sample: n_1. $n_2 < 30$, independent samples; population: normal distribution; standard deviations: unknown but equal
Test statistic: $$t = \dfrac{(\bar{x}_1 - \bar{x}_2) - (\mu_1 - \mu_2)}{\sqrt{\dfrac{(n_1 - 1)s_1^2 + (n_2 - 1)s_2^2}{n_1 + n_2 - 2}}\sqrt{\dfrac{1}{n_1} + \dfrac{1}{n_2}}}, \text{ degrees of freedom} = n_1 + n_2 - 2$$
Two-tailed hypothesis: Null hypothesis (H_0): $\mu_1 = \mu_2$ Alternative hypothesis (H_a): $\mu_1 \neq \mu_2$ Rejection criteria: $t < t_{\alpha/2, n1 + n2 - 2}$ or $t > t_{\alpha/2, n1 + n2 - 2}$
One-tailed hypothesis (upper/right tail):   Null hypothesis (H_0): $\mu_1 \leq \mu_2$ Alternative hypothesis (H_a): $\mu_1 > \mu_2$ Rejection criteria: $t > t_{\alpha, n1 + n2 - 2}$
One-tailed hypothesis (lower/left tail):   Null hypothesis (H_0): $\mu_1 \geq \mu_2$ Alternative hypothesis (H_a): $\mu_1 < \mu_2$ Rejection criteria: $t < t_{\alpha, n1 + n2 - 2}$

Table B.8   Paired *t*-test.

Paired *t*-test
Type: Mean
Assumptions: Sample: $n \leq 30$, sample sizes are equal; population type: normally distributed; standard deviations: approximately equal
Test statistic: $t = \dfrac{\bar{d}}{s_d/\sqrt{n}}$, where $\bar{d}$ is the mean of the differences between the two samples and s_d is the standard deviations of the differences and degrees of freedom $n - 1$
Two-tailed hypothesis: Null hypothesis (H_0): $\mu_d = \mu_0$ Alternative hypothesis (H_a): $\mu_d \neq \mu_0$ Rejection criteria: $t < -t_{\alpha/2,n-1}$ or $t > t_{\alpha/2,n-1}$
One-tailed hypothesis (upper/right tail):   Null hypothesis (H_0): $\mu_d \leq \mu_0$ Alternative hypothesis (H_a): $\mu_d > \mu_0$ Rejection criteria: $t > t_{\alpha,n1+n2-2}$
One-tailed hypothesis (lower/left tail):   Null hypothesis (H_0): $\mu_d \geq \mu_0$ Alternative hypothesis (H_a): $\mu_d < \mu_0$ Rejection criteria: $t < t_{\alpha,n-1}$

Table B.9   One-sample proportions.

One-sample hypothesis tests for proportion with any population and σ is not relevant
Type: Proportion
Assumptions: Sample size: $np \geq 5$, $nq \geq 5$; population type: binomial distribution; σ: not relevant
Test statistic: $$Z = \frac{(p - p_0)}{\sqrt{\dfrac{p_0(1 - p_0)}{n}}},$$ where p is the sample proportion (equal occurrences / opportunities) and p_0 is a hypothesized value for the proportion
Two-tailed hypothesis: Null hypothesis (H_0): $p = p_0$ Alternative hypothesis (H_a): $p \neq p_0$ Rejection criteria: $t < t_{\alpha/2, n-1}$ or $t > t_{\alpha/2, n-1}$
One-tailed hypothesis (upper/right tail): Null hypothesis (H_0): $p \leq p_0$ Alternative hypothesis (H_a): $p > p_0$ Rejection criteria: $Z > Z_\alpha$
One-tailed hypothesis (lower/left tail): Null hypothesis (H_0): $p \geq p_0$ Alternative hypothesis (H_a): $p < p_0$ Rejection criteria: $Z < Z_\alpha$

Table B.10 Two-sample proportions.

Two-sample hypothesis tests for proportion with any population and σ is not relevant
Type: Proportion
Assumptions: Sample size: $np \geq 5$, $nq \geq 5$; population type: binomial distribution; σ: not relevant
Test statistic: $$Z = \frac{(p_1 - p_2)}{\sqrt{\hat{p}(1-\hat{p})\left(\frac{1}{n_1} + \frac{1}{n_2}\right)}}, \text{ with } \hat{p} = \frac{n_1 p_1 + n_2 p_2}{n_1 + n_2}$$ where p_1 is the proportion for sample one and p_2 is the proportion for sample two
Two-tailed hypothesis: Null hypothesis (H_0): $p_1 - p_2 = 0$ Alternative hypothesis (H_a): $p_1 - p_2 > 0$ Rejection criteria: $Z < Z_{\alpha/2}$ or $Z > Z_{\alpha/2}$
One-tailed hypothesis (upper/right tail): Null hypothesis (H_0): $p_1 - p_2 \leq 0$ Alternative hypothesis (H_a): $p_1 - p_2 > p_0$ Rejection criteria: $Z > Z_{\alpha}$
One-tailed hypothesis (lower/left tail): Null hypothesis (H_0): $p_1 - p_2 \geq 0$ Alternative hypothesis (H_a): $p_1 - p_2 < 0$ Rejection criteria: $Z < Z_{\alpha}$

Table B.11 One-sample confidence interval.

Type: Mean
Assumptions: $n \geq 30$; population type: any shape; σ: known or $n < 30$; population type: normal distribution; σ: known
Test statistic: $$\bar{x} \pm Z_{\alpha/2} * \frac{\sigma}{\sqrt{n}}, \text{ degrees of freedom} = n$$

Table B.12 Confidence interval for mean with unknown standard deviation and $n \geq 30$.

Type: Mean
Assumptions: Sample size: $n \geq 30$; population type: any shape; σ: unknown
Test statistic: $$\bar{x} \pm Z_{\alpha/2} * \frac{s}{\sqrt{n}}$$

Table B.13 Confidence interval for mean with unknown standard deviation and $n < 30$.

Type: Mean
Assumptions: Sample size: $n < 30$; population type: normal distribution; σ: unknown
Test statistic: $$\bar{x} \pm t_{\alpha/2} * \frac{s}{\sqrt{n}}, \text{ degrees of freedom} = n - 1$$

Table B.14 Sample size for proportion tests.

Sample size determination: $$n = \left(\frac{Z_{\alpha/2}}{E}\right)^2 \hat{p}(1 - \hat{p}),$$ where E is the margin of error in percent and $\hat{p}$ is the estimated proportion
Confidence interval for proportions: $$p \pm Z_{\alpha/2} * \sqrt{\frac{\hat{p}(1 - \hat{p})}{n}}$$

Table B.15   Chi-square: one-sample variance test.

Type: Variance
Assumptions: Population type: normal distribution
Test statistic: $\chi^2 = (n - 1)\dfrac{S^2}{\sigma_0^2}$, degrees of freedom $= n - 1$
Two-tailed hypothesis: Null hypothesis (H_0): $\sigma = \sigma_0$ Alternative hypothesis (H_a): $\sigma \neq \sigma_0$ Rejection criteria: $\chi^2 < \chi^2_{n-1,1-\alpha/2}$ or $\chi^2 > \chi^2_{n-1,1-\alpha/2}$
One-tailed hypothesis (upper/right tail): Null hypothesis (H_0): $\sigma \leq \sigma_0$ Alternative hypothesis (H_a): $\sigma > \sigma_0$ Rejection criteria: $\chi^2 > \chi^2_{n-1,\alpha}$
One-tailed hypothesis (lower/left tail): Null hypothesis (H_0): $\sigma \geq \sigma_0$ Alternative hypothesis (H_a): $\sigma < \sigma_0$ Rejection criteria: $\chi^2 < \chi^2_{n-1,1-\alpha}$

Table B.16 *F*-test: two-sample variance test.

Type: Variance
Assumptions: Population type: both populations are normally distributed
Test statistic: $F = \dfrac{s_1^2}{s_2^2}$, degrees of freedom $= n_1 - 1, n_2 - 1$
Two-tailed hypothesis: Null hypothesis (H_0): $\sigma_1^2 = \sigma_2^2$ Alternative hypothesis (H_a): $\sigma_1^2 \neq \sigma_2^2$ Rejection criteria: $F > F_{\alpha/2,n1-1,n2-2}$ or $F < F_{1-\alpha/2,n1-1,n2-1}$
One-tailed hypothesis (upper/right tail): Null hypothesis (H_0): $\sigma_1^2 \leq \sigma_2^2$ Alternative hypothesis (H_a): $\sigma_1^2 > \sigma_2^2$ Rejection criteria: $F > F_{\alpha,n1-1,n2-1}$
One-tailed hypothesis (lower/left tail): Null hypothesis (H_0): $\sigma_1^2 \geq \sigma_2^2$ Alternative hypothesis (H_a): $\sigma_1^2 < \sigma_2^2$ Rejection criteria: $F_0 < F_{1-\alpha,n1-1,n2-1}$

Table B.17 Simple linear regression.

Type: Correlation
Test statistic:
$$r = \frac{\Sigma(x - \bar{x})(y - \bar{y})}{\sqrt{s\Sigma(x - \bar{x})^2}\sqrt{\Sigma(y - \bar{y})^2}} = \frac{\Sigma xy - (\Sigma x)(\Sigma y)/n}{\sqrt{\Sigma x^2 - (\Sigma x)^2/n}\sqrt{\Sigma y^2 - (\Sigma y)^2/n}}$$
$$SS_x = \Sigma x^2 - \frac{(\Sigma x)^2}{n}$$
$$SS_y = \Sigma y^2 - \frac{(\Sigma y)^2}{n}$$
$$SS_{xy} = \Sigma xy - \frac{(\Sigma x)(\Sigma y)}{n}$$
$$r = \frac{SS_{xy}}{\sqrt{SS_x}\sqrt{SS_y}}$$
The equation for the least squares line is $\hat{Y} = b_0 + b_1 x$, with $b_0 = \bar{y} - b_1\bar{x}$ and
$$b_1 = \frac{SCS_{xy}}{SS_x} \quad \text{and } \bar{x} = \frac{\Sigma x}{n} \text{ and } \bar{y} = \frac{\Sigma y}{n}$$
Hypothesis:
$H_0: \beta = 0$
$H_a: \beta \neq 0$

Table B.18 ANOVA.

Type: Comparison of multiple means
Assumptions: Samples must be independent of each other; population type: approximately normal distribution with equal variances
Test statistic: $SS \text{ (Factor)} = \left[\dfrac{T_1^2}{n_1} + \dfrac{T_2^2}{n_2} + \dfrac{T_3^2}{n_3} \right] - \dfrac{T^2}{n}$, with $T = T_1 + T_2 + T_3$ and $n = n_1 + n_2 + n_3$ $SS \text{ (Total)} = \Sigma x^2 - \dfrac{T^2}{n}$, with $\Sigma x^2 =$ the sum of the squared observations $SS \text{ (Error)} = SS \text{ (Total)} - SS \text{ (Factor)}$ $MS \text{ (Factor)} = \dfrac{SS \text{ (Factor)}}{k - 1}$ $MS \text{ (Error)} = \dfrac{SS \text{ (Error)}}{n - k} = S_P^2$
Hypothesis: Null hypothesis (H_0): $\mu_1 = \mu_2 = \mu_3$ Alternative hypothesis (H_a): not all μ's are equal Rejection criteria: $F > F_{.05, k-1, n-k}$

Appendix C
Statistical Tables

Table C.1 Standard normal Z score table (two tail or upper tail).

Z Value	0.00	0.01	0.02	0.03	0.04	0.05	0.06	0.07	0.08	0.09	Z Value
0.0	0.500000	0.503989	0.507978	0.511966	0.515953	0.519939	0.523922	0.527903	0.531881	0.535856	0.0
0.1	0.539828	0.543795	0.547758	0.551717	0.555670	0.559618	0.563559	0.567495	0.571424	0.575345	0.1
0.2	0.579260	0.583166	0.587064	0.590954	0.594835	0.598706	0.602568	0.606420	0.610261	0.614092	0.2
0.3	0.617911	0.621720	0.625516	0.629300	0.633072	0.636831	0.640576	0.644309	0.648027	0.651732	0.3
0.4	0.655422	0.659097	0.662757	0.666402	0.670031	0.673645	0.677242	0.680822	0.684386	0.687933	0.4
0.5	0.691462	0.694974	0.698468	0.701944	0.705401	0.708840	0.712260	0.715661	0.719043	0.722405	0.5
0.6	0.725747	0.729069	0.732371	0.735653	0.738914	0.742154	0.745373	0.748571	0.751748	0.754903	0.6
0.7	0.758036	0.761148	0.764238	0.767305	0.770350	0.773373	0.776373	0.779350	0.782305	0.785236	0.7
0.8	0.788145	0.791030	0.793892	0.796731	0.799546	0.802337	0.805105	0.807850	0.810570	0.813267	0.8
0.9	0.815940	0.818589	0.821214	0.823814	0.826391	0.828944	0.831472	0.833977	0.836457	0.838913	0.9
1.0	0.841345	0.843752	0.846136	0.848495	0.850830	0.853141	0.855428	0.857690	0.859929	0.862143	1.0
1.1	0.864334	0.866500	0.868643	0.870762	0.872857	0.874928	0.876976	0.879000	0.881000	0.882977	1.1
1.2	0.884930	0.886861	0.888768	0.890651	0.892512	0.894350	0.896165	0.897958	0.899727	0.901475	1.2
1.3	0.903200	0.904902	0.906582	0.908241	0.909877	0.911492	0.913085	0.914657	0.916207	0.917736	1.3
1.4	0.919243	0.920730	0.922196	0.923641	0.925066	0.926471	0.927855	0.929219	0.930563	0.931888	1.4
1.5	0.933193	0.934478	0.935745	0.936992	0.938220	0.939429	0.940620	0.941792	0.942947	0.944083	1.5
1.6	0.945201	0.946301	0.947384	0.948449	0.949497	0.950529	0.951543	0.952540	0.953521	0.954486	1.6
1.7	0.955435	0.956367	0.957284	0.958185	0.959070	0.959941	0.960796	0.961636	0.962462	0.963273	1.7
1.8	0.964070	0.964852	0.965620	0.966375	0.967116	0.967843	0.968557	0.969258	0.969946	0.970621	1.8
1.9	0.971283	0.971933	0.972571	0.973197	0.973810	0.974412	0.975002	0.975581	0.976148	0.976705	1.9
2.0	0.977250	0.977784	0.978308	0.978822	0.979325	0.979818	0.980301	0.980774	0.981237	0.981691	2.0

Z Value	0.00	0.01	0.02	0.03	0.04	0.05	0.06	0.07	0.08	0.09
2.1	0.982136	0.982571	0.982997	0.983414	0.983823	0.984222	0.984614	0.984997	0.985371	0.985738
2.2	0.986097	0.986447	0.986791	0.987126	0.987455	0.987776	0.988089	0.988396	0.988696	0.988989
2.3	0.989276	0.989556	0.989830	0.990097	0.990358	0.990613	0.990863	0.991106	0.991344	0.991576
2.4	0.991802	0.992024	0.992240	0.992451	0.992656	0.992857	0.993053	0.993244	0.993431	0.993613
2.5	0.993790	0.993963	0.994132	0.994297	0.994457	0.994614	0.994766	0.994915	0.995060	0.995201
2.6	0.995339	0.995473	0.995604	0.995731	0.995855	0.995975	0.996093	0.996207	0.996319	0.996427
2.7	0.996533	0.996636	0.996736	0.996833	0.996928	0.997020	0.997110	0.997197	0.997282	0.997365
2.8	0.997445	0.997523	0.997599	0.997673	0.997744	0.997814	0.997882	0.997948	0.998012	0.998074
2.9	0.998134	0.998193	0.998250	0.998305	0.998359	0.998411	0.998462	0.998511	0.998559	0.998605
3.0	0.998650	0.998694	0.998736	0.998777	0.998817	0.998856	0.998893	0.998930	0.998965	0.998999
3.1	0.999032	0.999065	0.999096	0.999126	0.999155	0.999184	0.999211	0.999238	0.999264	0.999289
3.2	0.999313	0.999336	0.999359	0.999381	0.999402	0.999423	0.999443	0.999462	0.999481	0.999499
3.3	0.999517	0.999534	0.999550	0.999566	0.999581	0.999596	0.999610	0.999624	0.999638	0.999651
3.4	0.999663	0.999675	0.999687	0.999698	0.999709	0.999720	0.999730	0.999740	0.999749	0.999758
3.5	0.999767	0.999776	0.999784	0.999792	0.999800	0.999807	0.999815	0.999822	0.999828	0.999835
3.6	0.999841	0.999847	0.999853	0.999858	0.999864	0.999869	0.999874	0.999879	0.999883	0.999888
3.7	0.999892	0.999896	0.999900	0.999904	0.999908	0.999912	0.999915	0.999918	0.999922	0.999925
3.8	0.999928	0.999931	0.999933	0.999936	0.999938	0.999941	0.999943	0.999946	0.999948	0.999950
3.9	0.999952	0.999954	0.999956	0.999958	0.999959	0.999961	0.999963	0.999964	0.999966	0.999967
4.0	0.999968	0.999970	0.999971	0.999972	0.999973	0.999974	0.999975	0.999976	0.999977	0.999978

Source: Based on a table available at www.cpkinfo.com and used with permission of Dean Christolear.

Table C.2 Standard normal Z score table (lower tail).

Z Value	0.00	0.01	0.02	0.03	0.04	0.05	0.06	0.07	0.08	0.09	Z Value
-4.0	0.000032	0.000030	0.000029	0.000028	0.000027	0.000026	0.000025	0.000024	0.000023	0.000022	-4.0
-3.9	0.000048	0.000046	0.000044	0.000042	0.000041	0.000039	0.000037	0.000036	0.000034	0.000033	-3.9
-3.8	0.000072	0.000069	0.000067	0.000064	0.000062	0.000059	0.000057	0.000054	0.000052	0.000050	-3.8
-3.7	0.000108	0.000104	0.000100	0.000096	0.000092	0.000088	0.000085	0.000082	0.000078	0.000075	-3.7
-3.6	0.000159	0.000153	0.000147	0.000142	0.000136	0.000131	0.000126	0.000121	0.000117	0.000112	-3.6
-3.5	0.000233	0.000224	0.000216	0.000208	0.000200	0.000193	0.000185	0.000178	0.000172	0.000165	-3.5
-3.4	0.000337	0.000325	0.000313	0.000302	0.000291	0.000280	0.000270	0.000260	0.000251	0.000242	-3.4
-3.3	0.000483	0.000466	0.000450	0.000434	0.000419	0.000404	0.000390	0.000376	0.000362	0.000349	-3.3
-3.2	0.000687	0.000664	0.000641	0.000619	0.000598	0.000577	0.000557	0.000538	0.000519	0.000501	-3.2
-3.1	0.000968	0.000935	0.000904	0.000874	0.000845	0.000816	0.000789	0.000762	0.000736	0.000711	-3.1
-3.0	0.001350	0.001306	0.001264	0.001223	0.001183	0.001144	0.001107	0.001070	0.001035	0.001001	-3.0
-2.9	0.001866	0.001807	0.001750	0.001695	0.001641	0.001589	0.001538	0.001489	0.001441	0.001395	-2.9
-2.8	0.002555	0.002477	0.002401	0.002327	0.002256	0.002186	0.002118	0.002052	0.001988	0.001926	-2.8
-2.7	0.003467	0.003364	0.003264	0.003167	0.003072	0.002980	0.002890	0.002803	0.002718	0.002635	-2.7
-2.6	0.004661	0.004527	0.004396	0.004269	0.004145	0.004025	0.003907	0.003793	0.003681	0.003573	-2.6
-2.5	0.006210	0.006037	0.005868	0.005703	0.005543	0.005386	0.005234	0.005085	0.004940	0.004799	-2.5
-2.4	0.008198	0.007976	0.007760	0.007549	0.007344	0.007143	0.006947	0.006756	0.006569	0.006387	-2.4
-2.3	0.010724	0.010444	0.010170	0.009903	0.009642	0.009387	0.009137	0.008894	0.008656	0.008424	-2.3
-2.2	0.013903	0.013553	0.013209	0.012874	0.012545	0.012224	0.011911	0.011604	0.011304	0.011011	-2.2
-2.1	0.017864	0.017429	0.017003	0.016586	0.016177	0.015778	0.015386	0.015003	0.014629	0.014262	-2.1
-2.0	0.022750	0.022216	0.021692	0.021178	0.020675	0.020182	0.019699	0.019226	0.018763	0.018309	-2.0

Z Value	0.00	0.01	0.02	0.03	0.04	0.05	0.06	0.07	0.08	0.09
-1.9	0.028717	0.028067	0.027429	0.026803	0.026190	0.025588	0.024998	0.024419	0.023852	0.023295
-1.8	0.035930	0.035148	0.034380	0.033625	0.032884	0.032157	0.031443	0.030742	0.030054	0.029379
-1.7	0.044565	0.043633	0.042716	0.041815	0.040930	0.040059	0.039204	0.038364	0.037538	0.036727
-1.6	0.054799	0.053699	0.052616	0.051551	0.050503	0.049471	0.048457	0.047460	0.046479	0.045514
-1.5	0.066807	0.065522	0.064255	0.063008	0.061780	0.060571	0.059380	0.058208	0.057053	0.055917
-1.4	0.080757	0.079270	0.077804	0.076359	0.074934	0.073529	0.072145	0.070781	0.069437	0.068112
-1.3	0.096800	0.095098	0.093418	0.091759	0.090123	0.088508	0.086915	0.085343	0.083793	0.082264
-1.2	0.115070	0.113139	0.111232	0.109349	0.107488	0.105650	0.103835	0.102042	0.100273	0.098525
-1.1	0.135666	0.133500	0.131357	0.129238	0.127143	0.125072	0.123024	0.121000	0.119000	0.117023
-1.0	0.158655	0.156248	0.153864	0.151505	0.149170	0.146859	0.144572	0.142310	0.140071	0.137857
-0.9	0.184060	0.181411	0.178786	0.176186	0.173609	0.171056	0.168528	0.166023	0.163543	0.161087
-0.8	0.211855	0.208970	0.206108	0.203269	0.200454	0.197663	0.194895	0.192150	0.189430	0.186733
-0.7	0.241964	0.238852	0.235762	0.232695	0.229650	0.226627	0.223627	0.220650	0.217695	0.214764
-0.6	0.274253	0.270931	0.267629	0.264347	0.261086	0.257846	0.254627	0.251429	0.248252	0.245097
-0.5	0.308538	0.305026	0.301532	0.298056	0.294599	0.291160	0.287740	0.284339	0.280957	0.277595
-0.4	0.344578	0.340903	0.337243	0.333598	0.329969	0.326355	0.322758	0.319178	0.315614	0.312067
-0.3	0.382089	0.378280	0.374484	0.370700	0.366928	0.363169	0.359424	0.355691	0.351973	0.348268
-0.2	0.420740	0.416834	0.412936	0.409046	0.405165	0.401294	0.397432	0.393580	0.389739	0.385908
-0.1	0.460172	0.456205	0.452242	0.448283	0.444330	0.440382	0.436441	0.432505	0.428576	0.424655
Z Value	0.00	0.01	0.02	0.03	0.04	0.05	0.06	0.07	0.08	0.09

Source: Based on a table available at www.cpkinfo.com and used with permission of Dean Christolear.

Table C.3 *t* Distribution (one-tailed probabilities).

		One-tailed probabilities		
		0.10	**0.05**	**0.01**
	1	3.078	6.314	31.82
	2	1.886	2.920	6.965
	3	1.638	2.353	4.541
	4	1.533	2.132	3.747
	5	1.476	2.015	3.365
	6	1.440	1.943	3.143
	7	1.415	1.895	2.998
	8	1.397	1.860	2.896
	9	1.383	1.833	2.821
	10	1.372	1.812	2.764
	11	1.363	1.796	2.718
	12	1.356	1.782	2.681
	13	1.350	1.771	2.650
Degrees of freedom	**14**	1.345	1.761	2.624
	15	1.341	1.753	2.602
	16	1.337	1.746	2.583
	17	1.333	1.740	2.567
	18	1.330	1.734	2.552
	19	1.328	1.729	2.539
	20	1.325	1.725	2.528
	21	1.323	1.721	2.518
	22	1.321	1.717	3.6,8
	23	1.318	1.714	2.500
	24	1.318	1.711	2.492
	25	1.316	1.708	2.485
	26	1.315	1.706	2.479
	27	1.314	1.703	2.473
	28	1.313	1.701	2.467
	29	1.311	1.699	2.462
	30	1.310	1.697	2.457
	32	1.309	1.694	2.449

Table C.3 *t* Distribution (one-tailed probabilities). (Continued)

		One-tailed probabilities		
		0.10	**0.05**	**0.01**
Degrees of freedom	**34**	1.307	1.691	2.441
	36	1.306	1.688	2.434
	38	1.304	1.686	2.426
	40	1.303	1.684	2.423
	42	1.302	1.682	2.418
	44	1.301	1.680	2.414
	46	1.300	1.679	2.410
	48	1.299	1.677	2.407
	50	1.299	1.676	2.403
	55	1.297	1.673	2.396
	60	1.296	1.671	2.390
	65	1.295	1.669	2.385
	70	1.294	1.667	2.381
	80	1.292	1.664	2.374
	100	1.290	1.660	2.364
	150	1.287	1.655	2.351
	200	1.286	1.653	2.345

Table C.4 *t* Distribution (two-tailed probabilities).

		Two-tailed probabilities		
		0.10	0.05	0.01
Degrees of freedom	1	6.314	12.71	63.66
	2	2.920	4.303	9.925
	3	2.353	3.182	5.841
	4	2.132	2.776	4.604
	5	2.015	2.571	4.032
	6	1.943	2.447	3.707
	7	1.895	2.365	3.499
	8	1.860	2.306	3.355
	9	1.833	2.262	3.250
	10	1.812	2.228	3.169
	11	1.796	2.201	3.106
	12	1.782	2.179	3.055
	13	1.771	2.160	3.012
	14	1.761	2.145	2.977
	15	1.753	2.131	2.947
	16	1.746	2.120	2.921
	17	1.740	2.110	2.898
	18	1.734	2.101	2.878
	19	1.729	2.093	2.861
	20	1.725	2.086	2.845
	21	1.721	2.080	2.831
	22	1.717	2.074	2.819
	23	1.714	2.069	2.807
	24	1.711	2.064	2.797
	25	1.708	2.060	2.787
	26	1.706	2.056	2.779
	27	1.703	2.052	2.771
	28	1.701	2.048	2.763
	29	1.699	2.045	2.756
	30	1.697	2.042	2.750
	32	1.694	2.037	2.738

Table C.4 *t* Distribution (two-tailed probabilities). (Continued)

		Two-tailed probabilities		
		0.10	**0.05**	**0.01**
	34	1.691	2.032	2.728
	36	1.688	2.028	2.719
	38	1.686	2.024	2.712
	40	1.684	2.021	2.704
	42	1.682	2.018	2.698
	44	1.680	2.015	2.692
	46	1.679	2.013	2.687
Degrees of freedom	**48**	1.677	2.011	2.682
	50	1.676	2.009	2.678
	55	1.673	2.004	2.668
	60	1.671	2.000	2.660
	65	1.669	1.997	2.654
	70	1.667	1.994	2.648
	80	1.664	1.990	2.639
	100	1.660	1.984	2.626
	150	1.655	1.976	2.609
	200	1.653	1.972	2.601

Table C.5   95% points for the *F* distribution (1–10 numerator degrees of freedom).

Denominator degrees of freedom	Numerator degrees of freedom									
	1	2	3	4	5	6	7	8	9	10
1	161	199	216	225	230	234	237	239	241	242
2	18.5	19.0	19.2	19.2	19.3	19.3	19.4	19.4	19.4	19.4
3	10.1	9.55	9.28	9.12	9.01	8.94	8.89	8.85	8.81	8.79
4	7.71	6.94	6.59	6.39	6.26	6.16	6.09	6.04	6.00	5.96
5	6.61	5.79	5.41	5.19	5.05	4.95	4.88	4.82	4.77	4.74
6	5.99	5.14	4.76	4.53	4.39	4.28	4.21	4.15	4.10	4.06
7	5.59	4.74	4.35	4.12	3.97	3.87	3.79	3.73	3.68	3.64
8	5.32	4.46	4.07	3.84	3.69	3.58	3.50	3.44	3.39	3.35
9	5.12	4.26	3.86	3.63	3.48	3.37	3.29	3.23	3.18	3.14
10	4.96	4.10	3.71	3.48	3.33	3.22	3.14	3.07	3.02	2.98
11	4.84	3.98	3.59	3.36	3.20	3.09	3.01	2.95	2.90	2.85
12	4.75	3.89	3.49	3.26	3.11	3.00	2.91	2.85	2.80	2.75
13	4.67	3.81	3.41	3.18	3.03	2.92	2.83	2.77	2.71	2.67
14	4.60	3.74	3.34	3.11	2.96	2.85	2.76	2.70	2.65	2.60
15	4.54	3.68	3.29	3.06	2.90	2.79	2.71	2.64	2.59	2.54
16	4.49	3.63	3.24	3.01	2.85	2.74	2.66	2.59	2.54	2.49
17	4.45	3.59	3.2	2.96	2.81	2.70	2.61	2.55	2.49	2.45
18	4.41	3.55	3.16	2.93	2.77	2.66	2.58	2.51	2.46	2.41
19	4.38	3.52	3.13	2.90	2.74	2.63	2.54	2.48	2.42	2.38

Denominator degrees of freedom

Numerator degrees of freedom

Denominator df	1	2	3	4	5	6	7	8	9	10
20	4.35	3.49	3.10	2.87	2.71	2.60	2.51	2.45	2.39	2.35
21	4.32	3.47	3.07	2.84	2.68	2.57	2.49	2.42	2.37	2.32
22	4.30	3.44	3.05	2.82	2.66	2.55	2.46	2.40	2.34	2.30
23	4.28	3.42	3.03	2.80	2.64	2.53	2.44	2.37	2.32	2.27
24	4.26	3.40	3.01	2.78	2.62	2.51	2.42	2.36	2.30	2.25
25	4.24	3.39	2.99	2.76	2.60	2.49	2.40	2.34	2.28	2.24
26	4.23	3.37	2.98	2.74	2.59	2.47	2.39	2.32	2.27	2.22
27	4.21	3.35	2.96	2.73	2.57	2.46	2.37	2.31	2.25	2.20
28	4.20	3.34	2.95	2.71	2.56	2.45	2.36	2.29	2.24	2.19
29	4.18	3.33	2.93	2.70	2.55	2.43	2.35	2.28	2.22	2.18
30	4.17	3.32	2.92	2.69	2.53	2.42	2.33	2.27	2.21	2.16
35	4.12	3.27	2.87	2.64	2.49	2.37	2.29	2.22	2.16	2.11
40	4.08	3.23	2.84	2.61	2.45	2.34	2.25	2.18	2.12	2.08
50	4.03	3.18	2.79	2.56	2.40	2.29	2.20	2.13	2.07	2.03
60	4.00	3.15	2.76	2.53	2.37	2.25	2.17	2.1	2.04	1.99
70	3.98	3.13	2.74	2.50	2.35	2.23	2.14	2.07	2.02	1.97
80	3.96	3.11	2.72	2.49	2.33	2.21	2.13	2.06	2.00	1.95
100	3.94	3.09	2.70	2.46	2.31	2.19	2.10	2.03	1.97	1.93
150	3.90	3.06	2.66	2.43	2.27	2.16	2.07	2.00	1.94	1.89
300	3.87	3.03	2.63	2.40	2.24	2.13	2.04	1.97	1.91	1.86
1000	3.85	3.00	2.61	2.38	2.22	2.11	2.02	1.95	1.89	1.84

Denominator degrees of freedom

Table C.6 95% points for the *F* distribution (11–20 numerator degrees of freedom).

Denominator degrees of freedom	Numerator degrees of freedom									
	11	12	13	14	15	16	17	18	19	20
1	243	244	245	245	246	246	247	247	248	248
2	19.4	19.4	19.4	19.4	19.4	19.4	19.4	19.4	19.4	19.4
3	8.76	8.74	8.73	8.71	8.70	8.69	8.68	8.67	8.67	8.66
4	5.94	5.91	5.89	5.87	5.86	5.84	5.83	5.82	5.81	5.80
5	4.70	4.68	4.66	4.64	4.62	4.60	4.59	4.58	4.57	4.56
6	4.03	4.00	3.98	3.96	3.94	3.92	3.91	3.90	3.88	3.87
7	3.60	3.57	3.55	3.53	3.51	3.49	3.48	3.47	3.46	3.44
8	3.31	3.28	3.26	3.24	3.22	3.20	3.19	3.17	3.16	3.15
9	3.10	3.07	3.05	3.03	3.01	2.99	2.97	2.96	2.95	2.94
10	2.94	2.91	2.89	2.86	2.85	2.83	2.81	2.80	2.79	2.77
11	2.82	2.79	2.76	2.74	2.72	2.70	2.69	2.67	2.66	2.65
12	2.72	2.69	2.66	2.64	2.62	2.60	2.58	2.57	2.56	2.54
13	2.63	2.60	2.58	2.55	2.53	2.51	2.50	2.48	2.47	2.46
14	2.57	2.53	2.51	2.48	2.46	2.44	2.43	2.41	2.40	2.39
15	2.51	2.48	2.45	2.42	2.40	2.38	2.37	2.35	2.34	2.33
16	2.46	2.42	2.40	2.37	2.35	2.33	2.32	2.30	2.29	2.28
17	2.41	2.38	2.35	2.33	2.31	2.29	2.27	2.26	2.24	2.23
18	2.37	2.34	2.31	2.29	2.27	2.25	2.23	2.22	2.20	2.19
19	2.34	2.31	2.28	2.26	2.23	2.21	2.20	2.18	2.17	2.16

Denominator degrees of freedom

Numerator degrees of freedom

Denominator df \ Numerator df	11	12	13	14	15	16	17	18	19	20
20	2.31	2.28	2.25	2.22	2.20	2.18	2.17	2.15	2.14	2.12
21	2.28	2.25	2.22	2.20	2.18	2.16	2.14	2.12	2.11	2.10
22	2.26	2.23	2.20	2.17	2.15	2.13	2.11	2.10	2.08	2.07
23	2.24	2.20	2.18	2.15	2.13	2.11	2.09	2.08	2.06	2.05
24	2.22	2.18	2.15	2.13	2.11	2.09	2.07	2.05	2.04	2.03
25	2.20	2.16	2.14	2.11	2.09	2.07	2.05	2.04	2.02	2.01
26	2.18	2.15	2.12	2.09	2.07	2.05	2.03	2.02	2.00	1.99
27	2.17	2.13	2.10	2.08	2.06	2.04	2.02	2.00	1.99	1.97
28	2.15	2.12	2.09	2.06	2.04	2.02	2.00	1.99	1.97	1.96
29	2.14	2.10	2.08	2.05	2.03	2.01	1.99	1.97	1.96	1.94
30	2.13	2.09	2.06	2.04	2.01	1.99	1.98	1.96	1.95	1.93
35	2.07	2.04	2.01	1.99	1.96	1.94	1.92	1.91	1.89	1.88
40	2.04	2.00	1.97	1.95	1.92	1.90	1.89	1.87	1.85	1.84
50	1.99	1.95	1.92	1.89	1.87	1.85	1.83	1.81	1.80	1.78
60	1.95	1.92	1.89	1.86	1.84	1.82	1.80	1.78	1.76	1.75
70	1.93	1.89	1.86	1.84	1.81	1.79	1.77	1.75	1.74	1.72
80	1.91	1.88	1.84	1.82	1.79	1.77	1.75	1.73	1.72	1.70
100	1.89	1.85	1.82	1.79	1.77	1.75	1.73	1.71	1.69	1.68
150	1.85	1.82	1.79	1.76	1.73	1.71	1.69	1.67	1.66	1.64
300	1.82	1.78	1.75	1.72	1.70	1.68	1.66	1.64	1.62	1.61
1000	1.80	1.76	1.73	1.70	1.68	1.65	1.63	1.61	1.60	1.58

Denominator degrees of freedom

Table C.7   95% points for the *F* distribution (21–30 numerator degrees of freedom).

Denominator degrees of freedom	Numerator degrees of freedom									
	21	**22**	**23**	**24**	**25**	**26**	**27**	**28**	**29**	**30**
1	248	249	249	249	249	249	250	250	250	250
2	19.4	19.5	19.5	19.5	19.5	19.5	19.5	19.5	19.5	19.5
3	8.65	8.65	8.64	8.64	8.63	8.63	8.63	8.62	8.62	8.62
4	5.79	5.79	5.78	5.77	5.77	5.76	5.76	5.75	5.75	5.75
5	4.55	4.54	4.53	4.53	4.52	4.52	4.51	4.50	4.50	4.50
6	3.86	3.86	3.85	3.84	3.83	3.83	3.82	3.82	3.81	3.81
7	3.43	3.43	3.42	3.41	3.40	3.40	3.39	3.39	3.38	3.38
8	3.14	3.13	3.12	3.12	3.11	3.10	3.10	3.09	3.08	3.08
9	2.93	2.92	2.91	2.90	2.89	2.89	2.88	2.87	2.87	2.86
10	2.76	2.75	2.75	2.74	2.73	2.72	2.72	2.71	2.70	2.70
11	2.64	2.63	2.62	2.61	2.60	2.59	2.59	2.58	2.58	2.57
12	2.53	2.52	2.51	2.51	2.50	2.49	2.48	2.48	2.47	2.47
13	2.45	2.44	2.43	2.42	2.41	2.41	2.40	2.39	2.39	2.38
14	2.38	2.37	2.36	2.35	2.34	2.33	2.33	2.32	2.31	2.31
15	2.32	2.31	2.30	2.29	2.28	2.27	2.27	2.26	2.25	2.25
16	2.26	2.25	2.24	2.24	2.23	2.22	2.21	2.21	2.20	2.19
17	2.22	2.21	2.20	2.19	2.18	2.17	2.17	2.16	2.15	2.15
18	2.18	2.17	2.16	2.15	2.14	2.13	2.13	2.12	2.11	2.11
19	2.14	2.13	2.12	2.11	2.11	2.10	2.09	2.08	2.08	2.07

Denominator degrees of freedom (table rows) / **Numerator degrees of freedom** (table columns)

Denominator	21	22	23	24	25	26	27	28	29	30
20	2.11	2.10	2.09	2.08	2.07	2.07	2.06	2.05	2.05	2.04
21	2.08	2.07	2.06	2.05	2.05	2.04	2.03	2.02	2.02	2.01
22	2.06	2.05	2.04	2.03	2.02	2.01	2.00	2.00	1.99	1.98
23	2.04	2.02	2.01	2.01	2.00	1.99	1.98	1.97	1.97	1.96
24	2.01	2.00	1.99	1.98	1.97	1.97	1.96	1.95	1.95	1.94
25	2.00	1.98	1.97	1.96	1.96	1.95	1.94	1.93	1.93	1.92
26	1.98	1.97	1.96	1.95	1.94	1.93	1.92	1.91	1.91	1.90
27	1.96	1.95	1.94	1.93	1.92	1.91	1.90	1.90	1.89	1.88
28	1.95	1.93	1.92	1.91	1.91	1.90	1.89	1.88	1.88	1.87
29	1.93	1.92	1.91	1.90	1.89	1.88	1.88	1.87	1.86	1.85
30	1.92	1.91	1.90	1.89	1.88	1.87	1.86	1.85	1.85	1.84
35	1.87	1.85	1.84	1.83	1.82	1.82	1.81	1.80	1.79	1.79
40	1.83	1.81	1.80	1.79	1.78	1.77	1.77	1.76	1.75	1.74
50	1.77	1.76	1.75	1.74	1.73	1.72	1.71	1.70	1.69	1.69
60	1.73	1.72	1.71	1.70	1.69	1.68	1.67	1.66	1.66	1.65
70	1.71	1.70	1.68	1.67	1.66	1.65	1.65	1.64	1.63	1.62
80	1.69	1.68	1.67	1.65	1.64	1.63	1.63	1.62	1.61	1.60
100	1.66	1.65	1.64	1.63	1.62	1.61	1.60	1.59	1.58	1.57
150	1.63	1.61	1.60	1.59	1.58	1.57	1.56	1.55	1.54	1.54
300	1.59	1.58	1.57	1.55	1.54	1.53	1.52	1.51	1.51	1.50
1000	1.57	1.55	1.54	1.53	1.52	1.51	1.50	1.49	1.48	1.47

Table C.8 95% points for the *F* distribution (31–40 numerator degrees of freedom).

	Numerator degrees of freedom									
Denominator degrees of freedom	31	32	33	34	35	36	37	38	39	40
1	250	250	250	251	251	251	251	251	251	251
2	19.5	19.5	19.5	19.5	19.5	19.5	19.5	19.5	19.5	19.5
3	8.61	8.61	8.61	8.61	8.60	8.60	8.60	8.60	8.60	8.59
4	5.74	5.74	5.74	5.73	5.73	5.73	5.72	5.72	5.72	5.72
5	4.49	4.49	4.48	4.48	4.48	4.47	4.47	4.47	4.47	4.46
6	3.80	3.80	3.80	3.79	3.79	3.79	3.78	3.78	3.78	3.77
7	3.37	3.37	3.36	3.36	3.36	3.35	3.35	3.35	3.34	3.34
8	3.07	3.07	3.07	3.06	3.06	3.06	3.05	3.05	3.05	3.04
9	2.86	2.85	2.85	2.85	2.84	2.84	2.84	2.83	2.83	2.83
10	2.69	2.69	2.69	2.68	2.68	2.67	2.67	2.67	2.66	2.66
11	2.57	2.56	2.56	2.55	2.55	2.54	2.54	2.54	2.53	2.53
12	2.46	2.46	2.45	2.45	2.44	2.44	2.44	2.43	2.43	2.43
13	2.38	2.37	2.37	2.36	2.36	2.35	2.35	2.35	2.34	2.34
14	2.30	2.30	2.29	2.29	2.28	2.28	2.28	2.27	2.27	2.27
15	2.24	2.24	2.23	2.23	2.22	2.22	2.21	2.21	2.21	2.2
16	2.19	2.18	2.18	2.17	2.17	2.17	2.16	2.16	2.15	2.15
17	2.14	2.14	2.13	2.13	2.12	2.12	2.11	2.11	2.11	2.10
18	2.10	2.10	2.09	2.09	2.08	2.08	2.07	2.07	2.07	2.06
19	2.07	2.06	2.06	2.05	2.05	2.04	2.04	2.03	2.03	2.03

Denominator degrees of freedom

Denominator df	Numerator degrees of freedom									
	40	**39**	**38**	**37**	**36**	**35**	**34**	**33**	**32**	**31**
20	1.99	2.00	2.00	2.01	2.01	2.01	2.02	2.02	2.03	2.03
21	1.96	1.97	1.97	1.98	1.98	1.98	1.99	1.99	2.00	2.00
22	1.94	1.94	1.95	1.95	1.95	1.96	1.96	1.97	1.97	1.98
23	1.91	1.92	1.92	1.93	1.93	1.93	1.94	1.94	1.95	1.95
24	1.89	1.90	1.09	1.90	1.91	1.91	1.92	1.92	1.93	1.93
25	1.87	1.88	1.88	1.88	1.89	1.89	1.90	1.90	1.91	1.91
26	1.85	1.86	1.86	1.87	1.87	1.87	1.88	1.88	1.89	1.89
27	1.84	1.84	1.84	1.85	1.85	1.86	1.86	1.87	1.87	1.88
28	1.82	1.82	1.83	1.83	1.84	1.84	1.85	1.85	1.86	1.86
29	1.81	1.81	1.81	1.82	1.82	1.83	1.83	1.84	1.84	1.85
30	1.79	1.80	1.80	1.80	1.81	1.81	1.82	1.82	1.83	1.83
35	1.74	1.74	1.74	1.75	1.75	1.76	1.76	1.77	1.77	1.78
40	1.69	1.70	1.70	1.71	1.71	1.72	1.72	1.73	1.73	1.74
50	1.63	1.64	1.64	1.65	1.65	1.66	1.66	1.67	1.67	1.68
60	1.59	1.60	1.60	1.61	1.61	1.62	1.62	1.63	1.64	1.64
70	1.57	1.57	1.58	1.58	1.59	1.59	1.60	1.60	1.61	1.62
80	1.54	1.55	1.55	1.56	1.56	1.57	1.58	1.58	1.59	1.59
100	1.52	1.52	1.52	1.53	1.54	1.54	1.55	1.55	1.56	1.57
150	1.48	1.48	1.49	1.49	1.50	1.50	1.51	1.51	1.52	1.53
300	1.43	1.44	1.45	1.45	1.46	1.46	1.47	1.48	1.48	1.49
1000	1.41	1.41	1.42	1.42	1.43	1.43	1.44	1.45	1.46	1.46

Numerator degrees of freedom

Denominator degrees of freedom

Table C.9   95% points for the *F* distribution (45–1000 numerator degrees of freedom).

Denominator degrees of freedom	Numerator degrees of freedom									
	45	50	60	70	80	100	120	150	300	1000
1	251	252	252	252	253	253	253	253	254	254
2	19.5	19.5	19.5	19.5	19.5	19.5	19.5	19.5	19.5	19.5
3	8.59	8.58	8.57	8.57	8.56	8.55	8.55	8.54	8.54	8.53
4	5.71	5.70	5.69	5.68	5.67	5.66	5.66	5.65	5.64	5.63
5	4.45	4.44	4.43	4.42	4.41	4.41	4.40	4.39	4.38	4.37
6	3.76	3.75	3.74	3.73	3.72	3.71	3.70	3.70	3.68	3.67
7	3.33	3.32	3.30	3.29	3.29	3.27	3.27	3.26	3.24	3.23
8	3.03	3.02	3.01	2.99	2.99	2.97	2.97	2.96	2.94	2.93
9	2.81	2.80	2.79	2.78	2.77	2.76	2.75	2.74	2.72	2.71
10	2.65	2.64	2.62	2.61	2.60	2.59	2.58	2.57	2.55	2.54
11	2.52	2.51	2.49	2.48	2.47	2.46	2.45	2.44	2.42	2.41
12	2.41	2.40	2.38	2.37	2.36	2.35	2.34	2.33	2.31	2.30
13	2.33	2.31	2.30	2.28	2.27	2.26	2.25	2.24	2.23	2.21
14	2.25	2.24	2.22	2.21	2.20	2.19	2.18	2.17	2.15	2.14
15	2.19	2.18	2.16	2.15	2.14	2.12	2.11	2.10	2.09	2.07
16	2.14	2.12	2.11	2.09	2.08	2.07	2.06	2.05	2.03	2.02
17	2.09	2.08	2.06	2.05	2.03	2.02	2.01	2.00	1.98	1.97
18	2.05	2.04	2.02	2.00	1.99	1.98	1.97	1.96	1.94	1.92
19	2.01	2.00	1.98	1.97	1.96	1.94	1.93	1.92	1.90	1.88

Denominator degrees of freedom

Numerator degrees of freedom

Denominator df	1000	300	150	120	100	80	70	60	50	45
20	1.85	1.86	1.89	1.90	1.91	1.92	1.93	1.95	1.97	1.98
21	1.82	1.83	1.86	1.87	1.88	1.89	1.90	1.92	1.94	1.95
22	1.79	1.81	1.83	1.84	1.85	1.86	1.88	1.89	1.91	1.92
23	1.76	1.78	1.80	1.81	1.82	1.84	1.85	1.86	1.88	1.90
24	1.74	1.76	1.78	1.79	1.80	1.82	1.83	1.84	1.86	1.88
25	1.72	1.73	1.76	1.77	1.78	1.80	1.81	1.82	1.84	1.86
26	1.70	1.71	1.74	1.75	1.76	1.78	1.79	1.80	1.82	1.84
27	1.68	1.70	1.72	1.73	1.74	1.76	1.77	1.79	1.81	1.82
28	1.66	1.68	1.70	1.71	1.73	1.74	1.75	1.77	1.79	1.80
29	1.65	1.66	1.69	1.70	1.71	1.73	1.74	1.75	1.77	1.79
30	1.63	1.65	1.67	1.68	1.70	1.71	1.72	1.74	1.76	1.77
35	1.57	1.58	1.61	1.62	1.63	1.65	1.66	1.68	1.70	1.72
40	1.52	1.54	1.56	1.58	1.59	1.61	1.62	1.64	1.66	1.67
50	1.45	1.47	1.50	1.51	1.52	1.54	1.56	1.58	1.60	1.61
60	1.40	1.42	1.45	1.47	1.48	1.50	1.52	1.53	1.56	1.57
70	1.36	1.39	1.42	1.44	1.45	1.47	1.49	1.50	1.53	1.55
80	1.34	1.36	1.39	1.41	1.43	1.45	1.46	1.48	1.51	1.52
100	1.30	1.32	1.36	1.38	1.39	1.41	1.43	1.45	1.48	1.49
150	1.24	1.27	1.31	1.33	1.34	1.37	1.39	1.41	1.44	1.45
300	1.17	1.21	1.26	1.28	1.30	1.32	1.34	1.36	1.39	1.41
1000	1.11	1.16	1.22	1.24	1.26	1.29	1.31	1.33	1.36	1.38

Denominator degrees of freedom

Table C.10 Critical points of the chi-square distribution.

d.f.	0.005	0.01	0.025	0.05	0.1	0.25	0.5	0.75	0.9	0.95	0.975	0.99	0.995
1	3.90E-05	0.00016	0.00098	0.0039	0.0158	0.1	0.455	1.32	2.71	3.84	5.02	6.63	7.88
2	0.01	0.0201	0.0506	0.103	0.211	0.58	1.39	2.77	4.61	5.99	7.38	9.21	10.6
3	0.0717	0.115	0.216	0.352	0.584	1.21	2.37	4.11	6.25	7.81	9.35	11.3	12.8
4	0.207	0.297	0.484	0.711	1.06	1.92	3.36	5.39	7.78	9.49	11.1	13.3	14.9
5	0.412	0.554	0.831	1.15	1.61	2.67	4.35	6.63	9.24	11.1	12.8	15.1	16.7
6	0.676	0.872	1.24	1.64	2.2	3.45	5.35	7.84	10.6	12.6	14.4	16.8	18.5
7	0.989	1.24	1.69	2.17	2.83	4.25	6.35	9.04	12.0	14.1	16.0	18.5	20.3
8	1.34	1.65	2.18	2.73	3.49	5.07	7.34	10.2	13.4	15.5	17.5	20.1	22.0
9	1.73	2.09	2.70	3.33	4.17	5.9	8.34	11.4	14.7	16.9	19.0	21.7	23.6
10	2.16	2.56	3.25	3.94	4.87	6.74	9.34	12.5	16.0	18.3	20.5	23.2	25.2
11	2.60	3.05	3.82	4.57	5.58	7.58	10.3	13.7	17.3	19.7	21.9	24.7	26.8
12	3.07	3.57	4.40	5.23	6.30	8.44	11.3	14.8	18.5	21.0	23.3	26.2	28.3
13	3.57	4.11	5.01	5.89	7.04	9.3	12.3	16.0	19.8	22.4	24.7	27.7	29.8
14	4.07	4.66	5.63	6.57	7.79	10.2	13.3	17.1	21.1	23.7	26.1	29.1	31.3
15	4.60	5.23	6.26	7.26	8.55	11.0	14.3	18.2	22.3	25.0	27.5	30.6	32.8
16	5.14	5.81	6.91	7.96	9.31	11.9	15.3	19.4	23.5	26.3	28.8	32.0	34.3
17	5.70	6.41	7.56	8.67	10.1	12.8	16.3	20.5	24.8	27.6	30.2	33.4	35.7
18	6.26	7.01	8.23	9.39	10.9	13.7	17.3	21.6	26.0	28.9	31.5	34.8	37.2
19	6.84	7.63	8.91	10.1	11.7	14.6	18.3	22.7	27.2	30.1	32.9	36.2	38.6
20	7.43	8.26	9.59	10.9	12.4	15.5	19.3	23.8	28.4	31.4	34.2	37.6	40.0
21	8.03	8.90	10.3	11.6	13.2	16.3	20.3	24.9	29.6	32.7	35.5	38.9	41.4
22	8.64	9.54	11.0	12.3	14.0	17.2	21.3	26.0	30.8	33.9	36.8	40.3	42.8

Cumulative probability

d.f.	0.005	0.01	0.025	0.05	0.1	0.25	0.5	0.75	0.9	0.95	0.975	0.99	0.995
23	9.26	10.2	11.7	13.1	14.8	18.1	22.3	27.1	32.0	35.2	38.1	41.6	44.2
24	9.89	10.9	12.4	13.8	15.7	19.0	23.3	28.2	33.2	36.4	39.4	43.0	45.6
25	10.5	11.5	13.1	14.6	16.5	19.9	24.3	29.3	34.4	37.7	40.6	44.3	46.9
26	11.2	12.2	13.8	15.4	17.3	20.8	25.3	30.4	35.6	38.9	41.9	45.6	48.3
27	11.8	12.9	14.6	16.2	18.1	21.7	26.3	31.5	36.7	40.1	43.2	47.0	49.6
28	12.5	13.6	15.3	16.9	18.9	22.7	27.3	32.6	37.9	41.3	44.5	48.3	51.0
29	13.1	14.3	16.0	17.7	19.8	23.6	28.3	33.7	39.1	42.6	45.7	49.6	52.3
30	13.8	15.0	16.8	18.5	20.6	24.5	29.3	34.8	40.3	43.8	47.0	50.9	53.7
31	14.5	15.7	17.5	19.3	21.4	25.4	30.3	35.9	41.4	45.0	48.2	52.2	55.0
32	15.1	16.4	18.3	20.1	22.3	26.3	31.3	37.0	42.6	46.2	49.5	53.5	56.3
33	15.8	17.1	19.0	20.9	23.1	27.2	32.3	38.1	43.7	47.4	50.7	54.8	57.6
34	16.5	17.8	19.8	21.7	24.0	28.1	33.3	39.1	44.9	48.6	52.0	56.1	59.0
35	17.2	18.5	20.6	22.5	24.8	29.1	34.3	40.2	46.1	49.8	53.2	57.3	60.3
36	17.9	19.2	21.3	23.3	25.6	30.0	35.3	41.3	47.2	51.0	54.4	58.6	61.6
37	18.6	20.0	22.1	24.1	26.5	30.9	36.3	42.4	48.4	52.2	55.7	59.9	62.9
38	19.3	20.7	22.9	24.9	27.3	31.8	37.3	43.5	49.5	53.4	56.9	61.2	64.2
39	20.0	21.4	23.7	25.7	28.2	32.7	38.3	44.5	50.7	54.6	58.1	62.4	65.5
40	20.7	22.2	24.4	26.5	29.1	33.7	39.3	45.6	51.8	55.8	59.3	63.7	66.8
41	21.4	22.9	25.2	27.3	29.9	34.6	40.3	46.7	52.9	56.9	60.6	65.0	68.1
42	22.1	23.7	26.0	28.1	30.8	35.5	41.3	47.8	54.1	58.1	61.8	66.2	69.3
43	22.9	24.4	26.8	29.0	31.6	36.4	42.3	48.8	55.2	59.3	63.0	67.5	70.6
44	23.6	25.1	27.6	29.8	32.5	37.4	43.3	49.9	56.4	60.5	64.2	68.7	71.9
45	24.3	25.9	28.4	30.6	33.4	38.3	44.3	51	57.5	61.7	65.4	70.0	73.2

Cumulative probability

Appendix D
Statistical Process Control Constants

Table D.1

Subgroup size n	A_2	D_2	D_3	D_4	A_3	C_4	B_3	B_4	E_2	A_2 for median charts
2	1.880	1.128	0	3.267	2.659	0.798	0	3.267	2.660	1.880
3	1.023	1.693	0	2.574	1.954	0.886	0	2.586	1.772	1.187
4	0.729	2.059	0	2.282	1.628	0.921	0	2.266	1.457	0.796
5	0.577	2.326	0	2.114	1.427	0.940	0	2.089	1.290	0.691
6	0.483	2.534	0	2.004	1.287	0.952	0.030	1.970	1.184	0.548
7	0.419	2.704	0.076	1.924	1.182	0.959	0.118	1.882	1.109	0.508
8	0.373	2.847	0.136	1.864	1.099	0.965	0.185	1.815	1.054	0.433
9	0.337	2.970	0.184	1.816	1.932	0.969	0.239	1.761	1.010	0.412
10	0.308	3.078	0.223	1.777	0.975	0.973	0.284	1.716	0.975	0.362

Reproduced from Connie M. Borror, ed., *The Certified Quality Engineer Handbook*, 3rd ed. (Milwaukee, WI: ASQ Quality Press, 2009).

Appendix E
Minitab Quick Reference

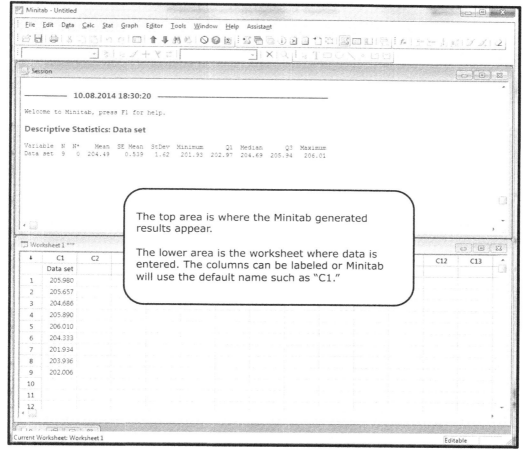

Figure E.1 Minitab screen.

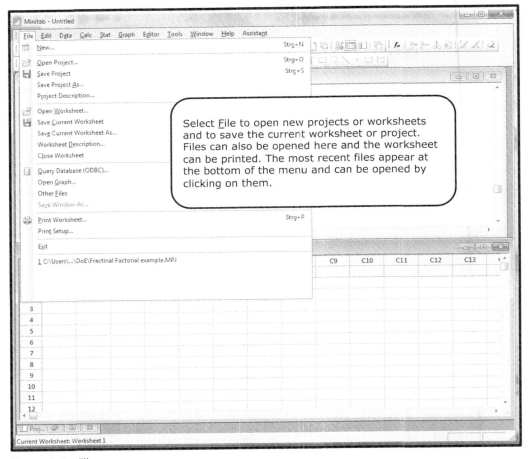

Figure E.2 File.

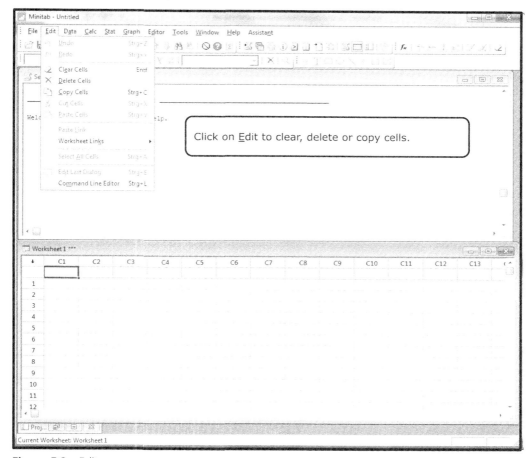

Figure E.3   Edit.

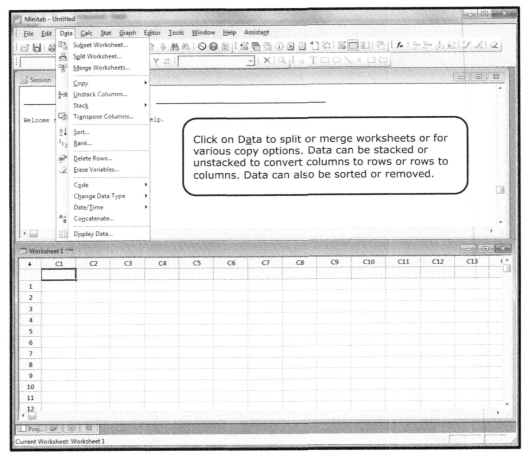

Figure E.4   Data.

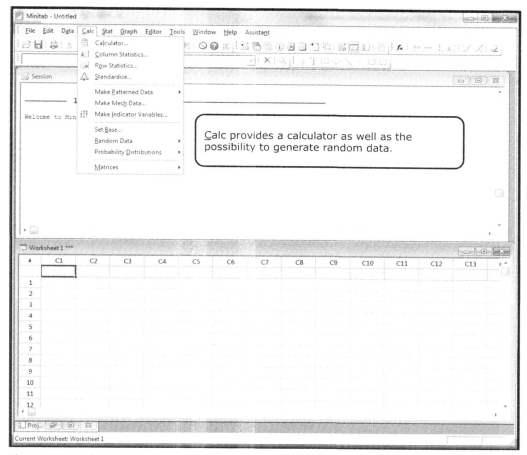

Figure E.5   Calc.

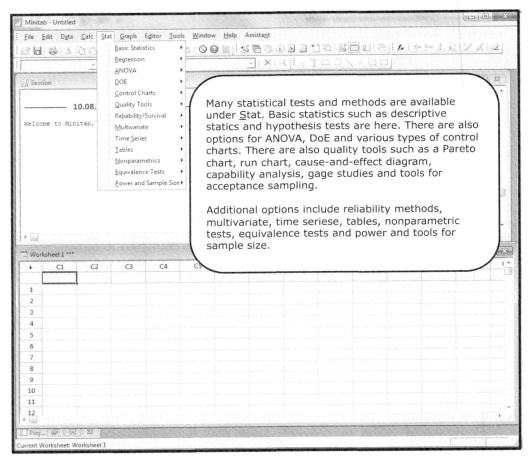

Figure E.6 Stat.

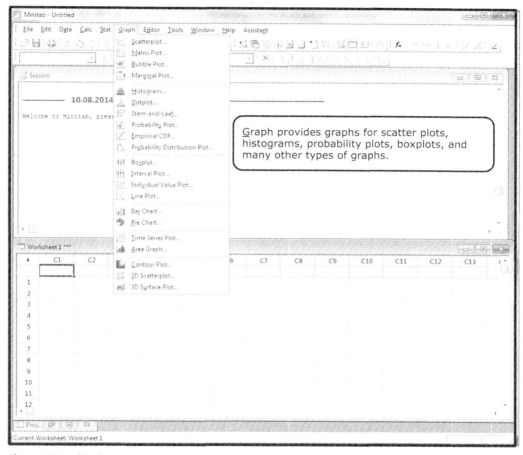

Graph provides graphs for scatter plots, histograms, probability plots, boxplots, and many other types of graphs.

Figure E.7 Graph.

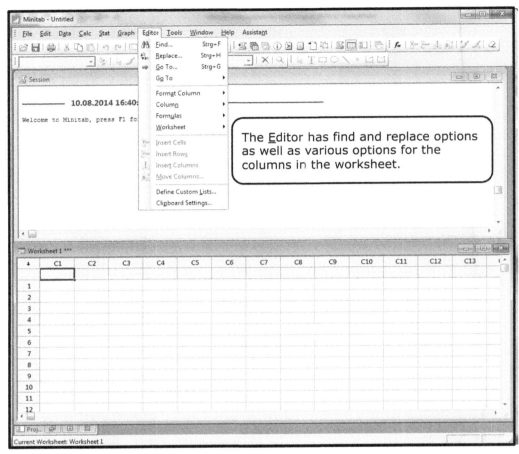

Figure E.8 Editor.

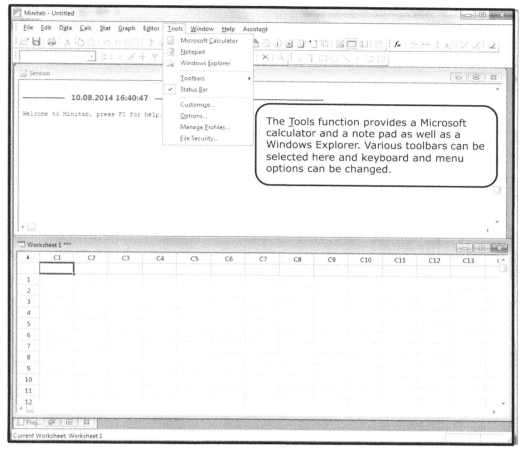

The Tools function provides a Microsoft calculator and a note pad as well as a Windows Explorer. Various toolbars can be selected here and keyboard and menu options can be changed.

Figure E.9   Tools.

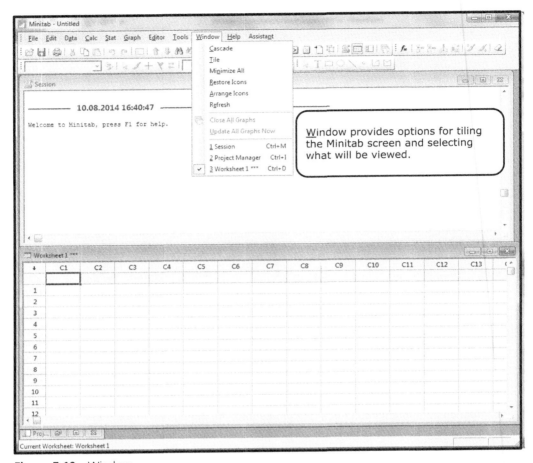

Figure E.10 Window.

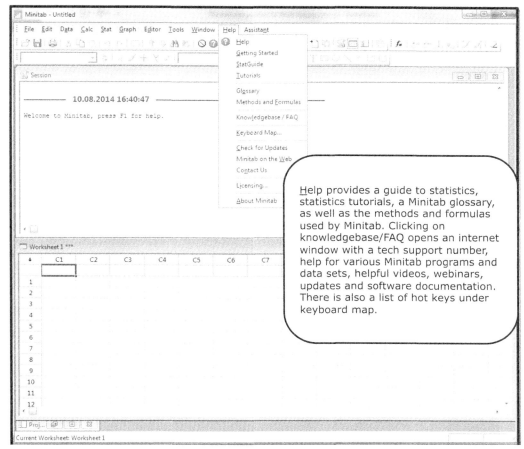

Figure E.11   Help.

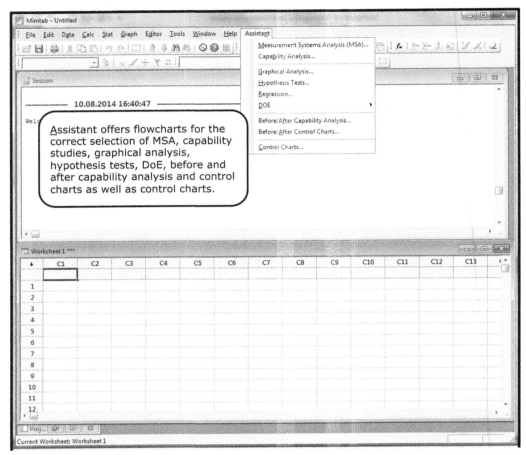

Figure E.12 Assistant.

Appendix F
Glossary

1 – α—See *confidence level*.

1 – β—The power of testing a hypothesis is $1 - \beta$. It is the probability of correctly rejecting the *null hypothesis, H_0*.[17]

α (alpha)—1. The maximum probability, or risk, of making a *type I error* when dealing with the *significance level* of a test. 2. The *probability* or risk of incorrectly deciding that a shift in the process mean has occurred when the process is unchanged (when referring to α in general or as the *p-value* obtained in a test). 3. α is usually designated as producer's risk.[8]

accuracy—The closeness of agreement between a test result or measurement result and the true value.[12]

action limits—The *control chart control limits* (for a process *in a state of statistical control*) beyond which there is a very high probability that a value is not due to chance. When a measured value lies beyond an action limit, appropriate corrective action should be taken on the process.

Example: Typical action limits for a $\bar{x}$ *chart* are $\pm 3\sigma$ (3 *standard deviations*).[12]

alias—An *effect* that is completely confounded with another effect due to the nature of the designed experiment. Aliases are the results of *confounding*, which may or may not be deliberate.[13]

alternative hypothesis, H1—A hypothesis that is accepted if the *null hypothesis* (H_0) is disproved.

Example: Consider the null hypothesis that the statistical model for a *population* is a *normal distribution*. The alternative hypothesis to this null hypothesis is that the statistical model of the population is *not* a normal distribution.

Note 1: The alternative hypothesis is a statement that contradicts the null hypothesis. The corresponding test statistic is used to decide between the null and alternative hypotheses.

Note 2: The alternative hypothesis can be denoted H_1, H_A, or H^A with no clear preference as long as the symbolism parallels the null hypothesis notation.[8,11]

Glossary entries reproduced and modified from the ASQ Statistical Division's *Glossary and Tables for Statistical Quality Control*, 4th ed. (Milwaukee, WI: ASQ Quality Press, 2005).

analysis of variance (ANOVA)—A technique to determine if there are statistically significant differences among group means by analyzing group *variances.* An ANOVA is an analysis technique that evaluates the importance of several factors of a set of data by subdividing the variation into component parts.

An analysis of variance table generally contains columns for:

- Source

- Degrees of freedom

- Sum of squares

- Mean square

- *F*-ratio or *F-test* statistic

- *p*-value

- Expected mean square

The basic assumptions are that the *effects* from the sources of *variation* are additive and that the *experimental errors* are independent, normally distributed, and have equal *variances.* ANOVA tests the *hypothesis* that the within-group variation is homogeneous and does not vary from group to group. The *null hypothesis* is that the group means are equal to each other. The *alternative hypothesis* is that at least one of the group means is different from the others.[13]

ANOVA—See *analysis of variance.*

arithmetic average—See *arithmetic mean.*

arithmetic mean—A calculation or estimation of the center of a set of values. It is a sum of the values divided by the number in the sum.

$$\bar{x} = \frac{1}{n} \sum_{i=1}^{n} x_i$$

where $\bar{x}$ is the arithmetic mean. The average of a set of n observed values is the sum of the observed values divided by n:

$$\bar{x} = \frac{x_1 + x_2 + \ldots + x_n}{n}$$

See also *population mean.*[5,11]

assignable cause—A specifically identified factor that contributes to *variation* and is feasible to detect and identify. Eliminating assignable causes so that the points plotted on a *control chart* remain within the *control limits* helps achieve a *state of statistical control.*

Note: Although assignable cause is sometimes considered synonymous with *special cause,* a special cause is assignable only when it is specifically identified.[12]

attribute—A countable or categorized quality *characteristic* that is *qualitative* rather than *quantitative* in nature. Attribute data come from *discrete, nominal,* or *ordinal scales.* Examples of attribute data are irregularities or flaws in a sample and results of pass/fail tests.[8,12]

attribute control chart—A *Shewhart control chart* where the measure plotted represents countable or categorized data.[12]

autocorrelation—The internal *correlation* between members of a series of observations ordered in time.

Example: Test values taken from hourly samples of a tank to which material is being continuously added. Each value represents material already in the tank plus the new material and contains material that was part of the previous sample.

Note: Autocorrelation can lead to misinterpretation of *runs* and trends in *control charts.*[12]

average—The central tendency. Common measures are the *mean, median,* or *mode* and the calculation depends on the type of distribution. If the term *average* is used without any descriptor, it is ordinarily the *arithmetic mean.*[21]

average run length (ARL)—The expected number of samples (or *subgroups*) plotted on a *control chart* up to and including the decision point that a *special cause* is present. The choice of ARL is a compromise between taking action when the process has not changed (ARL too small) or not taking action when a special cause is present (ARL too large).[12]

β (beta)—1. The maximum *probability,* or risk, of making a *type II error.* See comment on α *(alpha).* 2. The probability or risk of incorrectly deciding that a shift in the *process mean* has not occurred when the process has changed. 3. β is usually designated as *consumer's risk.*[2]

balanced design—A design where all *treatment* combinations have the same number of observations. If *replication* in a design exists, it would be balanced only if the replication was consistent across all the treatment combinations. In other words, the number of replicates of each treatment combination is the same.[19]

bias—Inaccuracy in a *measurement system* that occurs when the *mean* of the measurement result is consistently or systematically different than its *true value.*

bimodal—A probability distribution having two distinct statistical modes.[21]

binomial distribution—A two-parameter discrete distribution involving the *mean,* μ, and the *variance,* σ^2, of the variable x with *probability p* where p is a constant, $0 \le p \le 1$, and sample size n. Mean $= np$ and variance $= np(1-p)$.[21]

block—A collection of *experimental units* more homogeneous than the full set of experimental units. Blocks are usually selected to allow for *special causes,* in addition to those introduced as *factors* to be studied. These special causes may be avoidable within blocks, thus providing a more homogeneous experimental subspace.[8,13]

block effect—An *effect* resulting from a *block* in an *experimental design.* Existence of a block effect generally means that the method of blocking was appropriate.[21]

blocking—The method of including *blocks* in an experiment in order to broaden the applicability of the conclusions or to minimize the impact of selected *assignable causes.* The randomization of the experiment is restricted and occurs within blocks.[21]

box plot—Box plots, which are also called box-and-whisker plots, are particularly useful for showing the distributional characteristics of data. A box plot consists of a box, whiskers, and *outliers*. A line is drawn across the box at the *median*. By default, the bottom of the box is at the *first quartile* (Q1) and the top is at the *third quartile* (Q3) value. The whiskers are the lines that extend from the top and bottom of the box to the adjacent values. The adjacent values are the lowest and highest observations that are still inside the region defined by the following limits:

Lower limit: Q1 − 1.5 (Q3 − Q1)

Upper limit: Q3 + 1.5 (Q3 − Q1)

Outliers are points outside of the lower and upper limits and usually are plotted with asterisks (*).[14,16,21,23]

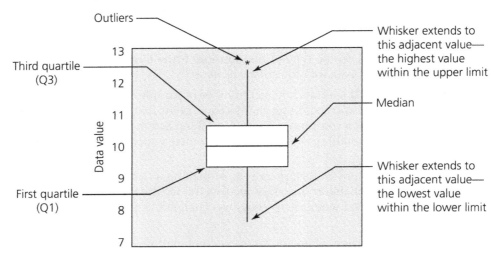

c (count)—The number of *events* (often *nonconformities*) of a given classification occurring in a *sample* of fixed size. A *nonconforming unit* may have more than one nonconformity.

Example: *Blemishes* per 100 meters of rubber hose. Counts are *attribute* data. See *c chart*.[2]

capability—The performance of a *process* demonstrated to be in a *state of statistical control*. See *process capability; process performance*.[21]

c **chart (count chart)**—An attribute control chart that uses *c* (count) or number of *events* as the plotted values where the opportunity for occurrence is fixed. Events, often *nonconformities*, of a particular type form the count. The fixed opportunity relates to samples of constant size or a fixed amount of material. Examples include flaws in each 100 square meters of fabric; errors in each 100 invoices; number of absentees per month.

Central line: $\bar{c}$

Control limits: $\bar{c} \pm 3\sqrt{\bar{c}}$

where $\bar{c}$ is the *arithmetic mean* of the number of events.

Note: If the *lower control limit* calculates ≤ 0, there is no lower control limit.

See also *u chart*, a count chart where the opportunity for occurrence is variable.[2,12]

center line—See *central line.*

center point—The *experimental runs* in a *factorial design* located at the center of the *design space.* In other words, all the *factor levels* are chosen to be halfway between the high and low settings. Center points can be used to test for *curvature* in the *response variable.* The center points can also be used to get an estimate of *standard deviation* or to test the significance of other *effects* where *replication* does not exist.[13]

central line—A line on a *control chart* representing the long-term *average* or a standard value of the *statistical measure* plotted. The central line calculation for each type of *control chart* is given under the specific control chart term; that is, for the *individuals control chart*, the central line is $\bar{x}$.[2]

chance cause—See *random cause.*

chance variation—See *random cause.*

chi-square distribution—See χ^2 *distribution.*

coefficient of determination (R^2)—A measure of the part of the *variance* for one *variable* that can be explained by its linear relationship with another variable (or variables). The coefficient of determination is the square of the correlation between the observed *y* values and the fitted *y* values, and is also the fraction of the variation in *y* that is explained by the fitted equation. It is a percentage between zero and 100 with higher values indicating a stronger degree of the combined linear relationship of several predictor variables $x1, x2, \ldots xp$ to the *response variable Y.* See *regression analysis.*[9,20]

common cause—See *random cause.*

confidence coefficient—A confidence interval is an estimate of the interval between two *statistics* that includes the *true value* of the *parameter* with some probability. This probability is called the *confidence level* of the estimate. Confidence levels typically used are 90 percent, 95 percent, and 99 percent. The interval either contains the parameter or it does not. See *t-confidence interval.*[17]

confidence level (confidence coefficient) ($1 - \alpha$)—1. The *probability* that the *confidence interval* described by a set of *confidence limits* actually includes the population parameter. 2. The probability that an interval about a sample *statistic* actually includes the population parameter.[18]

confidence limits—The endpoints of the interval about the sample *statistic* that is believed, with a specified *confidence level*, to include the population parameter.[9]

confounding—Indistinguishably combining an *effect* with other effects or *blocks.* When done deliberately, higher-order effects are systematically *aliased* so as to allow estimation of lower-order effects. Sometimes, confounding results from inadvertent changes to a design during the running of an experiment or poor planning of the design. This can diminish or even invalidate the effectiveness of the experiment.[13]

continuous distribution—A distribution where data is from a *continuous scale.* Examples of continuous scales include the *normal*, *t*, and *F* distributions.[21]

continuous scale—A scale with a continuum of possible values.

Note: A continuous scale can be transformed into a *discrete scale* by grouping values, but this leads to some loss of information.[12]

control chart—A chart that plots a *statistical measure* of a series of *samples* in a particular order to steer the *process* regarding that measure and to control and reduce variation.

Note 1: The order is usually time- or sample number order–based.

Note 2: The control chart operates most effectively when the measure is a process *characteristic* correlated with an ultimate product or service characteristic.[12]

control chart factor—A factor, usually varying with *sample size*, that converts specified statistics or parameters into *control limits* or a *central line* value.

Note: Common control chart factors for *Shewhart control charts* are *d2, A2, D3,* and *D4*.[2]

control limit—A line on a *control chart* used for judging the stability of a *process*.

Note 1: Control limits provide statistically determined boundaries for the *deviations* from the *central line* of the statistic plotted on a *Shewhart control chart* due to *random causes* alone.

Note 2: Control limits (with the exception of the *acceptance control chart*) are based on actual process data, not on *specification limits*.

Note 3: Other than points outside of *control limits*, out-of-control criteria can include *runs*, trends, cycles, periodicity, and unusual patterns within the control limits.

Note 4: The control limit calculation for each type of control chart is given under the specific control chart term; that is, for the *individuals control chart* under calculation of its control limits.[12]

control plan—A document describing the system elements to be applied to control *variation* of *processes*, products, and services in order to minimize deviation from their preferred values.[12]

correlation—Correlation measures the linear association between two variables. It is commonly measured by the *correlation coefficient, r*. See also *regression*.[21]

correlation coefficient (*r*)—A number between –1 and 1 that indicates the degree of linear relationship between two sets of numbers.

$$r^2 = \frac{(SS_{xy})^2}{(SS_x)(SS_y)} = \frac{n\Sigma xy - \Sigma y}{\sqrt{[n\Sigma x^2 - (\Sigma x)^2 - (\Sigma x)^2]}}$$

where *sx* is the *standard deviation* of *x*, *sy* is *standard deviation* of *y*, and *sxy* is the *covariance* of *x* and *y*. Correlation coefficients of –1 and +1 represent perfect linear agreement between two variables; *r* = 0 implies no linear relationship at all. If *r* is positive, *y* increases as *x* increases. In other words, if a *linear regression equation* were fit, the *linear regression coefficient* would be positive. If *r* is negative, *y* decreases as *x* increases. In other words, if a linear regression equation were fit, the linear regression coefficient would be negative.[9]

Cpk **(minimum process capability index)**—An index that represents the smaller of *CpkU* (*upper process capability index*) and *CpkL (lower process capability index)*.[12]

CpkL (**lower process capability index; CPL**)—An index describing *process capability* in relation to the lower *specification limit*.

$$\text{Cpk}_{\text{L}} = \frac{\mu - L}{3\sigma}$$

where μ = process average, L = lower specification limit, and 3σ = half of the *process capability*.[11,18]

CpkU (**upper process capability index; CPU**)—An index describing *process capability* in relation to the upper *specification limit*.

$$\text{Cpk}_{\text{U}} = \frac{U - \mu}{3\sigma}$$

where μ = process average, U = upper specification limit, and 3σ = half of the *process capability*.[7,12]

critical to quality (CTQ)—A *characteristic* of a product or service that is essential to ensure customer satisfaction.[21]

CTQ—See *critical to quality*.

defect—The nonfulfillment of a requirement related to an intended or specified use.

Note: The distinction between the concepts *defect* and *nonconformity* is important as it has legal connotations, particularly those associated with product liability issues. Consequently, the term *defect* should be used with extreme caution.[12]

defective (defective unit)—A *unit* with one or more *defects*.[12]

defects per million opportunities (DPMO)—The measure of capability for *discrete (attribute)* data found by dividing the number of *defects* by the opportunities for defects times a million. It allows for comparison of different types of product.[21]

defects per unit (DPU)—The measure of capability for *discrete (attribute)* data found by dividing the number of *defects* by the number of units.[21]

degrees of freedom (*v*, *d.f.*)—In general, the number of independent comparisons available to estimate a specific *parameter* that allows entry to certain statistical tables.[9]

dependent variable—See *response variable*.

design of experiments (DOE; DOX)—The arrangement in which an experimental program is to be conducted, including the selection of *factor* combinations and their *levels*.

Note: The purpose of designing an experiment is to provide the most efficient and economical methods of reaching valid and relevant conclusions from the experiment. The selection of the design is a function of many considerations, such as the type of questions to be answered, the applicability of the conclusions, the homogeneity of *experimental units*, the *randomization* scheme, and the cost to run the experiment. A properly designed experiment will permit simple interpretation of valid results.[13]

design resolution—See *resolution*.

design space—The multidimensional region of possible *treatment* combinations formed by the selected *factors* and their *levels*.

discrete distribution—A *probability distribution* where data are from a *discrete scale*. Examples of *discrete distributions* are *binomial* and *Poisson* distributions. *Attribute* data involve discrete distributions.[11]

discrete scale—A scale with only a set or sequence of distinct values.

Examples: Defects per unit, events in a given time period, types of defects, number of orders on a truck.[12]

discrimination—See *resolution*.

dispersion—A term synonymous with *variation*.

DOE—See *design of experiments*.

DOX—See *design of experiments*.

experiment space—See *design space*.

experimental design—See *design of experiments*.

experimental error—The variation in the *response variable* beyond that accounted for by the *factors*, *blocks*, or other assignable sources in the conduct of an experiment.[19]

experimental plan—The assignment of *treatments* to an *experimental unit* and the time order in which the *treatments* are to be applied.[19]

experimental run—A single performance of an experiment for a specific set of *treatment* combinations.[21]

experimental unit—The smallest entity receiving a particular *treatment*, subsequently yielding a value of the *response variable*.[13]

experimentation—See *design of experiments*.

explanatory variable—See *predictor variable*.

exploratory data analysis (EDA)—Isolates patterns and features of the data and reveals these forcefully to the analyst.[22]

F distribution—A *continuous distribution* that is a useful reference for assessing the ratio of independent *variances*.[12]

factor—A *predictor variable* that is varied with the intent of assessing its *effect* on a *response variable*.[19]

factor level—See *level*.

factorial design—An *experimental design* consisting of all possible *treatments* formed from two or more *factors*, each studied at two or more *levels*. When all combinations are run, the *interaction effects* as well as *main effects* can be estimated.[13]

first quartile (Q1 or lower quartile)—The portion of a distribution where one quarter of the data lies below. See *quartiles*.[21]

fixed factor—A *factor* that only has a limited number of *levels* that are of interest. In general, inference is not made to other levels of fixed factors not included in the experiment. For example, gender when used as a *factor* only has two possible *levels* of interest.[19,21]

fractional factorial design—An *experimental design* consisting of a subset (fraction) of the *factorial design*. Typically, the fraction is a simple proportion of the full set of possible *treatment* combinations. For example, half-fractions, quarter-fractions, and so forth are common. While fractional factorial designs require fewer runs, some degree of *confounding* occurs.[13]

F-test—A statistical test that uses the *F distribution*. It is most often used when dealing with a *hypothesis* related to the ratio of independent *variances*.

$$F = \frac{s_L^2}{s_S^2}$$

Where s_L^2 is the larger variance and s_S^2 is the smaller variance.

Fv1, v2—F-test statistic. See *F-test*.

gage R&R study—A type of *measurement system analysis* done to evaluate the performance of a test method or *measurement system*. Such a study quantifies the capabilities and limitations of a measurement instrument, often estimating its *repeatability* and *reproducibility*. It typically involves multiple operators measuring a series of measurement items multiple times.[19,21]

gaussian distribution—See *normal distribution*.

H0—See *null hypothesis*.

H1—See *alternative hypothesis*.

HA—See *alternative hypothesis*.

histogram—A plot of a *frequency distribution* in the form of rectangles (cells) whose bases are equal to the class interval and whose areas are proportional to the frequencies.[9]

Sample histogram

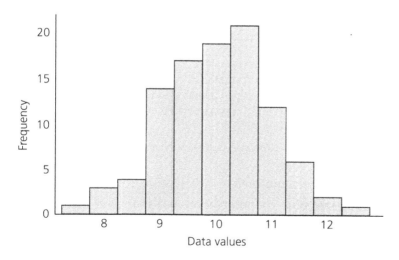

hypothesis—A statement about a population to be tested. See *null hypothesis; alternative hypothesis; hypothesis testing*.[11]

hypothesis testing—A statistical *hypothesis* is a conjecture about a *population parameter*. There are two statistical hypotheses for each situation—the *null hypothesis (H0)* and the *alternative hypothesis (H1)*. The null hypothesis proposes that there is no difference between the population of the sample and the specified population; the alternative hypothesis proposes that there is a difference between the sample and the specified population.[10]

I chart (individuals control chart)—See *individuals control chart.*

in-control process—A condition where the existence of *special causes* is no longer indicated by a *Shewhart control chart.* It indicates (within limits) a predictable and *stable process,* but it does not indicate that only *random causes* remain; nor does it imply that the distribution of the remaining values is normal (Gaussian).[18]

independent variable—See *predictor variable.*

individuals control chart (*x* chart; *I* chart)—A *variables control chart* with individual values plotted in the form of a *Shewhart control chart* for individuals and *subgroup* size $n = 1$.

Note 1: The *x* chart is widely used in the chemical industry because of cost of testing, test turnaround time, and the time interval between independent samples.

Note 2: This chart is usually accompanied by a *moving range chart,* commonly with $n = 2$.

Note 3: An individuals chart sacrifices the advantages of averaging subgroups (and the assumptions of the *normal distribution* central limit theorem) to minimize *random variation.*

Central line: $\bar{x}$ ($x0$ if standard given).

Control limits: $\bar{x} \pm E_3\overline{MR}$ or $\bar{x} \pm E_3\bar{s}$

(if standard given, $x_0 \pm 3\sigma_0$)
where $\bar{x}$ is the average of the individual variables

$$\bar{x} = \frac{x_1 + x_2 + \ldots + x_n}{n}$$

where R is the range, $\bar{s}$ is average sample standard deviation, $x0$ is the standard value for the average, and σ_0 is the standard value for the population standard deviation. (Use the formula with $\overline{MR}$ when the sample size is small; the formula with $\bar{s}$ when the sample is larger, generally >10 to 12.).[5,7,9,12]

inherent process variation—The variation in a *process* when the process is operating in a *state of statistical control.*[12]

input variable—A variable that can contribute to the variation in a *process.*[13]

inspection—A conformity evaluation by observation and judgment accompanied as appropriate by measurement, testing, or gauging.[12]

inspection by attributes—An *inspection* noting the presence, or absence, of the *characteristic*(s) in each of the items in the group under consideration and counting how many items do, or do not, possess the *characteristic*(s), or how many such events occur in the *item,* group, or opportunity space.

Note: When *inspection* is performed by simply noting whether the item is nonconforming or not, the inspection is termed inspection for *nonconforming* items. When inspection is performed by noting the number of nonconformities on each *unit,* the inspection is termed inspection for number of nonconformities.[12]

inspection by variables—An *inspection* measuring the magnitude(s) of the *characteristic*(s) of an item.[5]

interaction effect—The *effect* for which the apparent influence of one *factor* on the *response variable* depends upon one or more other factors. Existence of an *interaction effect* means that the factors cannot be changed independently of each other.[13]

interaction plot—The plot providing the *average* responses at the combinations of *levels* of two distinct *factors*.[13]

intercept—See *regression analysis*.

interquartile range (IQR)—The middle 50 percent of the data obtained by Q3 – Q1.[21]

IQR—See *interquartile range*.

kurtosis—A measure of peakedness or flattening of a distribution near its center in comparison to the *normal distribution*.

LCL—See *lower control limit*.

least squares, method of—A technique of estimating a *parameter* that minimizes the sum of the difference squared, where the difference is between the *observed value* and the predicted value *(residual)* derived from the *model*.

Note: *Experimental errors* associated with individual observations are assumed to be independent. The usual *analysis of variance, regression analysis,* and *analysis of covariance* are all based on the method of least squares.[13]

level—A potential setting, value, or assignment of a *factor* or the value of the *predictor variable*.[13]

level of significance—See *significance level*.

linear regression coefficients—The values associated with each *predictor variable* in a *linear regression equation*. They tell how the *response variable* changes with each unit increase in the *predictor variable*. See *regression analysis*.[9]

linear regression equation—A function that indicates the linear relationship between a set of *predictor variables* and a *response variable*. See *regression analysis*.[9]

linearity (general sense)—The degree to which a pair of variables follow a straight-line relationship. Linearity can be measured by the *correlation coefficient*.[21]

linearity (measurement system sense)—The difference in *bias* through the range of measurement. A measurement system that has good *linearity* will have a constant bias no matter the magnitude of measurement. If one views the relation between the observed measurement result on the y-axis and the *true value* on the x-axis, an ideal *measurement system* would have a line of slope equal to one.[19]

lower control limit (LCL)—The *control chart* control limit that defines the lower control boundary.[12]

lower specification limit (lower tolerance limit, L)—The *specification* or *tolerance limit* that defines the lower limiting value.[12]

µ (mu)—See *population mean*.

main effect—The influence of a single *factor* on the *mean* of a *response variable*.[13]

main effects plot—A plot giving the *average* responses at the various *levels* of individual *factors*.[13]

mean—See *arithmetic mean*.

means, tests for—Testing for *means* includes computing a *confidence interval* and hypothesis testing by comparing means to a *population* mean (known or unknown) or to other *sample* means. Which test to use is determined by whether *s* is known, whether the test involves one or two samples, and other factors.

If σ is known, the *z-confidence interval* and *z-test* apply. If σ is unknown, the *t-confidence interval* and *t-test* apply.[17]

measurement system—Everything that can introduce variability into the measurement *process*, such as equipment, operator, sampling methods, and accuracy.[21]

measurement systems analysis (MSA)—A statistical analysis of a *measurement system* to determine the variability, *bias, precision,* and *accuracy* of the measurement system(s). Such studies may also include the limit of detection, selectivity, *linearity,* and other *characteristics* of the system in order to determine the suitability of the measurement system for its intended purpose.

median—The value for which half the *data* is larger and half is smaller. The median provides an estimator that is insensitive to very extreme values in a data set, whereas the *average* is affected by extreme values.

Note: For an odd number of units, the median is the middle measurement; for an even number of units, the median is the average of the two middle measurements.[9,11]

method of least squares—See *least squares, method of.*

midrange—(Highest value + lowest value) / 2.[21]

mode—The most frequent value of a *variable*.[9]

model—The description relating a *response variable* to *predictor variable*(s) and including attendant assumptions.[13]

moving range chart (MR chart)—A *control chart* that plots the absolute difference between the current and previous value. It often accompanies an *individuals chart* or *moving average chart*. Note: The current observation replaces the oldest of the latest *n* + 1 observations.

Central line: $\bar{R}$

Control limits: UCL = $D_4\bar{R}$

$$LCL = D_3\bar{R}$$

where D4 and D3 are control chart factors.[7]

MR chart—See *moving range chart.*

MSA—See *measurement systems analysis.*

multiple linear regression—See *regression analysis.*

n—See *sample size.*

nominal scale—A scale with unordered, labeled categories, or a scale ordered by convention.

Examples: Type of defect, breed of dog, complaint category.

Note: It is possible to count by category, but not order or measure.[12]

nonconforming unit—A *unit* with one or more *nonconformities*.[12]

nonconformity—The nonfulfillment of a requirement. See notes under *defect*.[6]

normal distribution (Gaussian distribution)—A *continuous*, symmetrical *frequency distribution* that produces a bell-shaped curve. The location *parameter* (x-axis) is the *mean*, μ. The scale parameter, σ, is the *standard deviation* of the normal distribution. When measurements have a normal distribution, 68.26% of the values lie within plus or minus one standard deviation of the mean ($\mu \pm 1\sigma$); 95.44% lie within plus or minus two standard deviations of the mean ($\mu \pm 2\sigma$); while 99.73% lie within plus or minus three standard deviations of the mean ($\mu \pm 3\sigma$).[11]

np **(number of affected or categorized units)**—The total number of *units* in a *sample* in which an *event* of a given classification occurs. A unit (area of opportunity) is counted only once, even if several events of the same classification are encountered.

Note: In the quality field, the classification generally is number of *nonconforming units*.[2]

np **chart (number of categorized units control chart)**—An *attribute control chart* for number of *events* per unit where the opportunity is variable. (The *np* chart is for the number nonconforming whereas the *p chart* is for the proportion nonconforming.)

Note: Events of a particular type, for example, number of absentees or number of sales leads, form the count. In the quality field, events are often expressed as *nonconformities* and the variable opportunity relates to *subgroups* of variable size or variable amounts of material.

If a standard is given:

    Central line: *np*

    Control limits: $np \pm 3\sqrt{np(1-p)}$

where *np* is the standard value of the events per unit and *p* is the standard value of the fraction nonconforming.

If no standard is given:

    Central line: $\overline{np}$

    Control limits: $\overline{np} \pm 3\sqrt{\overline{np}(1-\overline{p})}$

where $\overline{np}$ is the average value of the events per unit and $-p$ is the average value of the fraction nonconforming.[8,12]

null hypothesis, H0—The *hypothesis* in *tests of significance* that there is no difference (null) between the *population* of the *sample* and the specified population (or between the populations associated with each sample). The null hypothesis can never be proved true, but it can be shown (with specified risks or error) to be untrue; that is, a difference exists between the populations. If it is not disproved, one assumes there is no adequate reason to doubt it is true, and the null hypothesis is accepted. If the null hypothesis is shown to be untrue, then the *alternative hypothesis* is accepted.

Example: In a random sample of independent random variables with the same *normal distribution* with unknown *mean* and unknown *standard deviation*, a typical null hypothesis for the mean, μ, is that the mean is less than or equal to a given value, μ_0. The hypothesis is written as: $H_0 = \mu \le \mu_0$.[8,11]

observed value—The particular value of a *characteristic* determined as a result of a test or measurement.

one-tailed test—A *hypothesis* test that involves only one of the tails of a distribution.

Example: We wish to reject the *null hypothesis H0* only if the true *mean* is larger than μ_c.

$$H_0: \mu = \mu_0$$
$$H_1: \mu < \mu_0$$

A one-tailed test is either right-tailed or left-tailed, depending on the direction of the inequality of the *alternative hypothesis*.[10,17]

ordinal scale—A scale with ordered, labeled categories.

Note 1: There is sometimes a blurred borderline between ordinal and *discrete scales*. When subjective opinion ratings such as excellent, very good, neutral, poor, and very poor are coded (as numbers 1–5), the apparent effect is conversion from an ordinal to a discrete scale. Such numbers should not be treated as ordinary numbers, however, because the distance between 1 and 2 may not be the same as between 2 and 3, or 3 and 4, and so forth. On the other hand, some categories that are ordered objectively according to magnitude, such as the Richter scale, which ranges from 0 to 8 according to the amount of energy release, could equally well be related to a discrete scale.

Note 2: Sometimes *nominal scales* are ordered by convention. An example is the blood groups A, B, and O, which are always stated in this order. The same is the case if different categories are denoted by single letters; they are then ordered by convention, according to the alphabet.[12]

out-of-control process—A *process* operating with the presence of *special causes*. See *in-control process*.[20]

outlier—An extremely high or an extremely low data value compared to the rest of the data values. Great caution must be used when trying to identify an outlier.

output variable—The variable representing the outcome of the *process*.[13]

p—The ratio of the number of *units* in which at least one *event* of a given classification occurs, to the total number of units. A unit is counted only once even if several events of the same classification are encountered within it. *P* can also be expressed as a percent.[2]

P—See *probability*.

parameter—A constant or coefficient describing some *characteristic* of a *population*. Examples: *standard deviation* and *mean*.[9]

parts per million (PPM or ppm)—A measurement that is expressed by dividing the data set into 1,000,000 or 106 equal groups.[21]

part-to-part variation—The variability of the data due to measurement items rather than the *measurement system*. This *variation* is typically estimated from the measurement items used in a study, but could be estimated from a *representative sample* of product.

p **chart**—See *proportion chart*.

Pearson's correlation coefficient—See *correlation coefficient*.

percentile—The division of the data set into 100 equal groups.[21]

pooled standard deviation—A *standard deviation* value resulting from some combination of individual standard deviation values. It is most often used when individual standard deviation values are similar in magnitude and can be denoted by sp.[21]

population—The entire set (totality) of *units*, quantity of material, or observations under consideration. A population may be real and finite, real and infinite, or completely hypothetical. See *sample*.[11,12]

population mean (μ)—The true *mean* of the *population*, represented by μ *(mu)*. The *sample mean*, $\bar{x}$, is a common estimator of the population mean.[21]

population standard deviation—See *standard deviation*.

population variance—See *variance*.

power—The equivalent to one minus the *probability* of a *type II error* $(1 - \beta)$. A higher power is associated with a higher probability of finding a statistically significant difference. Lack of power usually occurs with smaller sample sizes.[11]

power curve—The curve showing the relationship between the *probability* $(1 - \beta)$ of rejecting the *hypothesis* that a sample belongs to a given *population* with a given *characteristic*(s) and the actual population value of that characteristic(s).

Note: If b, the probability of accepting the hypothesis, is used instead of $(1 - \beta)$, the curve is called an *operating characteristic (OC) curve*.[2]

Pp **(process performance index)**—An index describing *process performance* in relation to specified *tolerance*

$$Pp = \frac{U - L}{6s}$$

s is used for *standard deviation* instead of σ since both *random* and *special causes* may be present.

Note: A *state of statistical control* is not required.[12]

Ppk **(minimum process performance index)**—The smaller of *upper process performance index* and *lower process performance index*.[12]

PpkL **(lower process performance index or PPL)**—An index describing *process performance* in relation to the *lower specification limit*. For a symmetrical normal distribution:

$$Ppk_{L} = \frac{\bar{x} - L}{3s}$$

where s is defined under *Pp*.[12]

PpkU **(upper process performance index or PPU)**—An index describing *process performance* in relation to the *upper specification limit*. For a symmetrical *normal distribution*:

$$Ppk_{U} = \frac{U - \bar{x}}{3s}$$

where s is defined under *Pp*.[12]

PPL—See *PpkL*.

PPM (or ppm)—See *parts per million*.

PPU—See *PpkU*.

precision—The closeness of agreement between randomly selected individual measurements or test results. See *repeatability; reproducibility*.[8]

predicted value—The prediction of future observations based on the formulated *model*.[21]

prediction interval—Similar to a *confidence interval*, it is an interval based on the *predicted value* that is likely to contain the values of future observations. It will be wider than the confidence interval because it contains bounds on individual observations rather than a bound on the *mean* of a group of observations.[21]

predictor variable—A *variable* that can contribute to the explanation of the outcome of an experiment.[19]

probability (*P*)—The chance of an *event* occurring.[21]

process average—See *population mean*.

process capability—The calculated inherent variability of a *characteristic* of a product. It represents the best performance of the *process* over a period of stable operations. Process capability is expressed as 6σ, where σ is the *sample standard deviation* (short-term component of variation) of the process under a *state of statistical control*.[7]

process capability index—A single-number assessment of ability to meet *specification limits* on the quality *characteristic*(s) of interest. The indices compare the variability of the characteristic to the specification limits. Three basic process capability indices are *Cp, Cpk*, and *Cpm*.

Note: Since there are many different types and variations of process capability indices, details are given under the symbol for the specific type of index.[15]

process control—*Process* management that is focused on fulfilling process requirements. Process control is the methodology for keeping a process within boundaries and minimizing the variation of a process.[8,12]

process performance—The statistical measure of the outcome of a *characteristic* from a *process* that may *not* have been demonstrated to be in a *state of statistical control*.

Note: Use this measure cautiously since it may contain a component of variability from *special causes* of unpredictable value. It differs from *process capability* because a state of statistical control is not required.[12]

process performance index—A single-number assessment of ability to meet *specification limits* on the quality *characteristic*(s) of interest. The indices compare the variability of the characteristic to the specification limits. Three basic process capability indices are *Pp, Ppk*, and *Ppm*.

Note: Since there are many different types and variations of process capability indices, details are given under the symbol for the specific type of index.[15]

process quality—A statistical measure of the quality of product from a given *process*. The measure may be an *attribute (qualitative)* or a *variable (quantitative)*. A common measure of process quality is the fraction or proportion of nonconforming units in the process.[21]

process variable—See *variable*.

proportion chart (*p* chart or percent categorized units control chart)—An *attribute control chart* for number of *units* of a given classification per total number of units in the *sample* expressed as either a proportion or percent.

Note 1: The classification often takes the form of *nonconforming units*.

Note 2: The *p* chart applies particularly when the sample size is variable.

Note 3: If the *upper control limit (UCL)* calculates ≥ 1 there is no UCL; or if the *lower control limit (LCL)* calculates ≤ 0, there is no LCL.

a. When the fraction nonconforming is known, or is a specified standard value:

Central line: *p*

Control limits: $p \pm 3\sqrt{\dfrac{p(1-p)}{n}}$

where *p* is known fraction nonconforming (or standard value).

Plotted value: $\hat{p}$

where $\hat{p}$ is the *sample* fraction nonconforming:

$\hat{p} = D/n$ where D is the number of product units that are nonconforming and *n* is the sample size.

b. When the fraction nonconforming is not known:

Central line: $\bar{p}$

Control limits: $\bar{p} \pm 3\sqrt{\dfrac{\bar{p}(1-\bar{p})}{n}}$

where –*p* is the *average* value of the fraction of the classification (often average percent nonconforming), and *n* is the total number of units. These limits are considered trial limits.

c. For variable sample size:

Central line: $\bar{p}$

Control limits: $\bar{p} \pm 3\sqrt{\dfrac{\bar{p}(1-\bar{p})}{n}}$

where *ni* is varying sample size.

Plotted value: $\hat{p}$

where $\hat{p}$ is the *sample* fraction nonconforming:
$\hat{p} = D_i/n$

where D_i is the number of nonconforming units in sample *i*, and *n* is the sample size.[8,12,17]

proportions, tests for—Tests for proportions include the *binomial distribution*.

The *standard deviation* for proportions is given by

$$s = \sqrt{\dfrac{p(1-p)}{n}}$$

where *p* is the *population* proportion and *n* is the *sample* size.[21]

p-value—The *probability* of observing the *test statistic* value or any other value at least as unfavorable to the *null hypothesis*.[11]

Q1—See *first quartile*.

Q2—See *second quartile*.

Q3—See *third quartile*.

quartiles—Division of a distribution into four groups, denoted by *Q1* (first quartile), *Q2* (second quartile), and *Q3* (third quartile). Note that *Q1* is the same as the 25th *percentile*, *Q2* is the same as the 50th percentile and the *median*, and *Q3* corresponds to the 75th percentile.[21]

r—See *correlation coefficient*.

R—See *range*.

$\bar{R}$ **(pronounced r-bar)**—The *average* range calculated from the set of *subgroup* ranges under consideration. See *range*.[21]

*R*²—See *coefficient of determination*.

random cause—The source of *process* variation that is inherent in a process over time. Also called *common cause* or *chance cause*.

Note: In a process subject only to random cause variation, the variation is predictable within statistically established limits.[12]

random sampling—A sampling where a *sample* of *n* sampling *units* is taken from a *population* in such a way that each of the possible combinations of *n* sampling units has a particular probability of being taken.[12]

random variation—Variation from *random causes*.[12]

randomization—The process used to assign *treatments* to *experimental units* so that each experimental unit has an equal chance of being assigned a particular *treatment*. Randomization validates the assumptions made in statistical analysis and prevents unknown biases from impacting the conclusions.[19]

randomized block design—An *experimental design* consisting of *b blocks* with *t treatments* assigned via *randomization* to the *experimental units* within each block. This is a method for controlling the variability of experimental units. For the *completely randomized design*, no stratification of the experimental units is made. In the randomized block design, the treatments are randomly allotted within each block; that is, the randomization is restricted.[13]

randomized block factorial design—A *factorial design* run in a *randomized block design* where each *block* includes a complete set of factorial combinations.[13]

range (*R*)—A measure of *dispersion*, which is the absolute difference between the highest and lowest *value* in a given *subgroup:* R = highest observed value – lowest observed value.[7,8]

range chart (*R* chart)—A *variables control chart* that plots the range of a *subgroup* to detect shifts in the subgroup range. See *range*.

Central line: $\bar{R}$

Upper control limit: $D_4\bar{R}$

Lower control limit: $D_3\bar{R}$

where $\bar{R}$ is the average range; *D3* and *D4* are factors.

Note: A range chart is used when the sample size is small; if the sample is larger (generally >10 to 12), the *s* chart should be used.[4,5]

rational subgroup—A *subgroup* wherein the variation is presumed to be only from *random causes.*[12]

R **chart**—See *range chart.*

regression—See *regression analysis.*

regression analysis—A technique that uses *predictor variable(s)* to predict the *variation* in a *response variable.* Regression analysis uses the method of *least squares* to determine the values of the *linear regression coefficients* and the corresponding *model.* It is particularly pertinent when the predictor variables are *continuous* and emphasis is on creating a predictive model. When some of the predictor variables are *discrete, analysis of variance* or *analysis of covariance* is likely a more appropriate method.

This resulting model can then test the resulting predictions for statistical significance against an appropriate *null hypothesis* model. The model also gives some sense of the degree of *linearity* present in the data. When only one predictor variable is used, regression analysis is often referred to as *simple linear regression.* A simple linear regression model commonly uses a *linear regression equation* expressed as $Y = \beta_0 + \beta_1 x + \varepsilon$, where Y is the response, x is the value of the predictor variable, β_0 and β_1 are the linear regression coefficients, and Σ is the random error term. β_0 is often called the *intercept* and β_1 is often called the *slope.*

When multiple predictor variables are used, regression is referred to as *multiple linear regression.* For example, a multiple linear regression model with three predictor variables commonly uses a linear regression equation expressed as $Y = \beta_0 + \beta_1 x_1 + \beta_2 x_2 + \beta_3 x_3 + \varepsilon$, where Y is the response, $x1$, $x2$, $x3$ are the values of the predictor variables, $\beta_0, \beta_1, \beta_2$, and β_3 are the linear regression coefficients, and ε is the random error term.

The random error terms in regression analysis are often assumed to be normally distributed with a constant *variance.* These assumptions can be readily checked through *residual analysis* or *residual plots.*[8,13,21]

See *regression coefficients, tests for.*

regression coefficients, tests for—A test of the individual *linear regression coefficients* to determine their significance in the *model.* These tests assume that the *response variable* is normally distributed for a fixed level of the *predictor variable,* the variability of the response variable is the same regardless of the value of the predictor variable, and that the predictor variable can be measured without error. See *regression analysis.*[21]

repeatability—*Precision* under conditions where independent measurement results are obtained with the same method on identical measurement items by the same operator using the same equipment within a short period of time.[12]

repeated measures—The measurement of a *response variable* more than once under similar conditions. Repeated measures allow one to determine the inherent variability in the *measurement system.* Also known as *duplication* or repetition.[21]

replicate—A single repetition of the experiment. See also *replication.*[13]

replication—Performance of an experiment more than once for a given set of *predictor variables.* Each of the repetitions of the experiment is called a *replicate.* Replication differs from *repeated measures* in that it is a repeat of the entire experiment for a given set of *predictor variables,* not just a repeat of measurements on the same experiment.

Note: Replication increases the precision of the estimates of the *effects* in an experiment. It is more effective when all elements contributing to the *experimental error* are included. In some cases replication may be limited to *repeated measures* under essentially the same conditions. In other cases, replication may be deliberately different, though similar, in order to make the results more general.[8,13]

reproducibility—*Precision* under conditions where independent measurement results are obtained with the same method on identical measurement items with different operators using different equipment.[12]

residual plot—A plot used in *residual analysis* to determine appropriateness of assumptions made by a statistical method. Common forms include a plot of the *residuals* versus the *observed values* or a plot of the residuals versus the *predicted values* from the fitted model.[21]

residuals—The difference between the observed result and the *predicted value* (estimated treatment response) based on empirically determined model.[13]

resolution—1. The smallest measurement increment that can be detected by the *measurement system*. 2. In the context of *experimental design*, resolution refers to the level of *confounding* in a *fractional factorial design*. For example in a resolution III design, the *main effects* are confounded with the two-way *interaction effects*.[13,21]

response variable—A variable representing the outcome of an experiment.[13]

run (control chart usage)—An uninterrupted sequence of occurrences of the same *attribute* or *event* in a series of observations, or a consecutive set of successively increasing (run up) or successively decreasing (run down) values in a series of *variable* measurements.

Note: In some *control chart* applications, a run might be considered a series of a specified number of points consecutively plotting above or below the *center line*, or five consecutive points, three of which fall outside warning limits.[2]

σ **(sigma)**—See *standard deviation*.

σ^2 **(sigma square)**—See *variance*.

$\sigma_{\bar{x}}$ **(sigma x-bar)**—The *standard deviation* (or *standard error*) of $\bar{x}$.[21]

$\hat{\sigma}$ **(sigma-hat)**—In general, any estimate of the *population standard deviation*. There are various ways to get this estimate depending on the particular application.[21]

s—See *standard deviation*.

s^2—See *variance*.

sample—A group of *units*, portions or material, or observations taken from a larger collection of units, quantity of material, or observations that serves to provide information that may be used for making a decision concerning the larger quantity (the *population*).

Note 1: The sample may be the actual units or material or the observations collected from them. The decision may or may not involve taking action on the units or material, or on the *process*.

Note 2: *Sampling plans* are schemes set up statistically in order to provide a sampling system with minimum bias.

Note 3: There are many different ways, random and nonrandom, to select a sample. In survey sampling, sampling units are often selected with a probability proportional to the size of a known variable, giving a biased sample.[9,12]

sample size (*n*)—The number of sampling *units* in a *sample*.

Note: In a multistage sample, the sample size is the total number of sampling *units* at the conclusion of the final stage of sampling.[12]

sample standard deviation—See *standard deviation*.

sample variance—See *variance*.

***s* chart**—See *standard deviation chart*.

second quartile (Q2)—The 50th *percentile* or the *median*. See *quartiles*.[21]

Shewhart control chart—A *control chart* with *Shewhart control limits* intended primarily to distinguish between variation due to *random causes* and variation due to *special causes*.[12]

Shewhart control limits—*Control limits* based on empirical evidence and economic considerations, placed about the *center line* at a distance of ± 3 *standard deviations* of the *statistic* under consideration and used to evaluate whether or not it is a *stable process*.[12]

signal—An indication on a *control chart* that a *process* is not *stable* or that a shift has occurred. Typical indicators are points outside *control limits*, *runs*, *trends*, cycles, patterns, and so on. See also *average run length*.[21]

significance level—The maximum *probability* of rejecting the *null hypothesis* when in fact it is true.

Note: The significance level is usually designated by α and should be set before beginning the test.[11]

significance tests—Significance tests are a method of deciding, with certain predetermined risks of error, (1) whether the *population* associated with a *sample* differs from the one specified; (2) whether the population associated with each of two samples differ; or (3) whether the populations associated with each of more than two samples differ. Significance testing is equivalent to the testing of *hypotheses*. Therefore, a clear statement of the *null hypothesis*, *alternative hypotheses*, and predetermined selection of a *confidence level* are required.

Note: The level of significance is the maximum *probability* of committing a *type I error*. This probability is symbolized by α; that is, P (type I error) = α. See *means, tests for*; *proportions, tests for*.[9,10]

simple linear regression—See *regression analysis*.

Six Sigma—A methodology that provides businesses with the tools to improve the capability of their business processes.[1]

skewness—A measure of symmetry about the *mean*. For a *normal distribution*, skewness is zero because the distribution is symmetric.[21]

slope—See *regression analysis*.

SPC—See *statistical process control*.

special cause—A source of process variation other than *inherent process variation*.

Note 1: Sometimes special cause is considered synonymous with *assignable cause*, but a special cause is assignable only when it is specifically identified.

Note 2: A special cause arises because of specific circumstances that are not always present. Therefore, in a process subject to special causes, the magnitude of the variation over time is unpredictable.[12]

specification limit(s)—The limiting value(s) stated for a *characteristic*.[12]

spread—A term sometimes synonymous with *variation* or *dispersion*.[21]

stable process—A *process* that is predictable within limits; a process that is subject only to *random causes*. (This is also known as a *state of statistical control*.)

Note 1: A stable process will generally behave as though the results are simple random samples from the same *population*.

Note 2: This state does not imply that the *random variation* is large or small, within or outside of *specification limits*, but rather that the variation is predictable using statistical techniques.

Note 3: The *process capability* of a stable process is usually improved by fundamental changes that reduce or remove some of the *random causes* present and/or adjusting the *mean* toward the *target* value.[18,19]

standard deviation—A measure of the *spread* of the *process* output or the spread of a sampling *statistic* from the process. When working with the *population*, the standard deviation is usually denoted by *s* (sigma). When working with a *sample* the standard deviation is usually denoted by *s*. They are calculated as

$$\sigma = \sqrt{\frac{1}{n}\Sigma(x-\mu)^2}$$

$$s = \sqrt{\frac{1}{n-1}\Sigma(x-\bar{x})^2}$$

where *n* is the number of data points in the sample or population, μ is the *population mean*, *x* is the observed value of the quality characteristic, and $\bar{x}$ is the sample mean. See *standard error*.

Note: Standard deviation can also be calculated by taking the square root of the *population variance* or *sample variance*.[11]

standard deviation chart (s chart)—A *variables control chart* of the *standard deviation* of the results within a *subgroup*. It replaces *range chart* for large subgroup samples (rule of thumb is subgroup size 10 to 12).

Central line: $\bar{s}$

Upper control limit: $B_4\bar{s}$

Lower control limit: $B_3\bar{s}$

where $\bar{s}$ is the average value of the standard deviation of the subgroups and *B3* and *B4* are factors. See *standard deviation*.[8,12]

standard error—The *standard deviation* of a *sample statistic* or estimator. When dealing with sample statistics, we either refer to the standard deviation of the sample statistic or to its standard error.[17]

standard error of predicted values—A measure of the variation of individual predicted values of the *dependent variable* about the *population* value for a given value of the *predictor variable.* This includes the variability of individuals about the sample regression line and the sample line about the population line. It measures the variability of individual observations and can be used to calculate a *prediction* interval.[9]

state of statistical control—See *stable process.*

statistic—A quantity calculated from a sample of observations, most often to form an estimate of some *population parameter.*[9]

statistical measure—A *statistic* or mathematical function of a statistic.[9]

statistical process control (SPC)—The use of statistical techniques such as *control charts* to reduce variation, increase knowledge about the *process,* and to steer the process in the desired way.

Note 1: SPC operates most efficiently by controlling variation of the process or in-process *characteristics* that correlate with a final product characteristic and/or by increasing the *robustness* of the process against this variation.

Note 2: A supplier's final product characteristic can be a process characteristic of the next downstream supplier's process.[12]

stratified sampling—A *sampling* such that portions of the *sample* are drawn from different strata and each stratum is sampled with at least one sampling *unit.*

Note 1: In some cases, the portions are specified proportions determined in advance, however, in post-stratified sampling the specified proportions would not be known in advance.

Note 2: Items from each stratum are often selected by random sampling.[12]

subgroup (control chart usage)—A group of data plotted as a single point on a control chart. See *rational subgroup.*[21]

target value—The preferred reference value of a *characteristic* stated in a specification.[12]

***t*-confidence interval for means (one-sample)**—When there is an unknown *mean* μ and unknown *variance* σ^2. For a *two-tailed test,* a 100(1 – σ)% two-sided confidence interval on the true mean:

$$\bar{x} - t_{\alpha/2,n-1}\frac{s}{\sqrt{n}} \le \mu \le \bar{x} + t_{\alpha/2,n-1}\frac{s}{\sqrt{n}}$$

where $t_{a/2,n-1}$ denotes the percentage point of the *t distribution* with *n* – 1 *degrees of freedom* such that $P\{t_{n-1} \le t_{a/2,n-1}\} = \sigma/2$.[17]

***t* distribution**—A theoretical distribution widely used in practice to evaluate the *sample mean* when the *population standard deviation* is estimated from the data. Also known as Student's distribution. It is similar in shape to the *normal distribution* with slightly longer tails. See *t-test.*[11]

third quartile (Q3 or upper quartile)—The portion of a distribution where one quarter of the data lies above it. See *quartiles.*[21]

tolerance limit—See *specification limit.*

treatment—The specific setting of *factor levels* for an *experimental unit*.[13]

t-test—A *test for significance* that uses the *t* distribution to compare a *sample statistic* to a hypothesized *population mean* or to compare two means. See *t-test (one sample)*.

Note: Testing the equality of the means of two normal populations with unknown but equal variances can be extended to the comparison of *k* population means. This test procedure is called *analysis of variance (ANOVA)*.[11]

t-test (one-sample)—$t = \dfrac{\bar{x} - \mu_0}{s/\sqrt{n}}$

where $\bar{x}$ is the mean of the data, μ_0 is the hypothesized *population mean*, *s* is the *sample standard deviation*, and *n* is the *sample* size. The degrees of freedom are $n - 1$.[10]

two-tailed test—A *hypothesis* test that involves tails of a distribution. Example: We wish to reject the *null hypothesis*, *H0*, if the true *mean* is within minimum and maximum (two tails) limits.[17]

$$H_0: \mu = \mu 0$$

$$H_1: \mu \neq \mu 0$$

type I error—The *probability* or risk of rejecting a *hypothesis* that is true. This probability is represented by α *(alpha)*. See diagram below.[12]

type II error—The *probability* or risk of accepting a *hypothesis* that is false. This probability is represented by β *(beta)*. See diagram below.[8]

	H_0 True	H_0 False
Do not reject H_0	Correct decision	Error Type II
Reject H_0	Error Type I	Correct decision

u (count per unit)—The *events* or events per *unit* where the opportunity is variable. More than one event may occur in a unit.

Note: For *u*, the opportunity is variable; for *c*, the opportunity for occurrence is fixed.[2]

u chart—An *attribute control chart* for number of *events* per unit where the opportunity is variable.

Note: Events of a particular type, for example, number of absentees or number of sales leads, form the count. In the quality field, events are often expressed as non-conformities and the variable opportunity relates to *subgroups* of variable size or variable amounts of material.

Central line: $\bar{u}$

Control limits: $\bar{u} \pm 3\sqrt{\pi/n}$

where $-u$ is the average number of events per unit and n is the total number of samples. $\bar{u}$ is calculated as $\bar{u} = \bar{c}/n$.

Note: If the *lower control limit (LCL)* calculates ≤ 0 there is no LCL.[8,12]

UCL—See *upper control limit.*

upper control limit (UCL)—The *control limit* on a *control chart* that defines the upper control boundary.[12]

upper quartile—See *third quartile.*

upper specification limit or upper tolerance limit (U)—The *specification limit* that defines the upper limiting value.[12]

variable (control chart usage)—A quality *characteristic* that is from a *continuous scale* and is *quantitative* in nature.[21]

variables control chart—A *Shewhart control chart* where the measure plotted represents data on a *continuous scale.*[12]

variance—A measure of the *variation* in the data. When working with the entire *population,* the population variance is used; when working with a *sample,* the sample variance is used. The population variance is based on the mean of the squared deviations from the *arithmetic mean* and is given by The sample variance is based on the squared deviations from the *arithmetic mean* divided by n – 1 and is given by

$$\sigma^2 = \frac{1}{n} \sum (x - \mu)^2$$

The sample variance is based on the squared deviations from the *arithmetic mean* divided by n – 1 and is given by

$$x^2 = \frac{1}{n} \sum (x - \bar{x})^2$$

where n is the number of data points in the sample or population, m is the *population mean,* x is the *observed value* of the quality characteristic, and $\bar{x}$ is the sample mean. See *standard deviation.*[21]

variances, tests for—A formal statistical test based on the *null hypothesis* that the *variances* of different groups are equal. Many times in *regression analysis* a formal test of variances is not done. Instead, *residual analysis* checks the assumption of equal variance across the values of the *response variable* in the *model.* For two variances, see *F-test.*[21]

variation—The difference between values of a *characteristic*. Variation can be measured and calculated in different ways, such as *range, standard deviation*, or *variance*. Also known as *dispersion* or *spread*.[12]

warning limits—There is a high *probability* that the *statistic* under consideration is in a *state of statistical control* when it is within the warning limits (generally 2σ) of a *control chart*. See *Shewhart control limits*.

Note: When the value of the statistic plotted lies outside a warning limit but within the *action limit*, increased supervision of the *process* to prespecified rules is generally required.[12]

χ^2 distribution—(pronounced chi-square) A positively skewed distribution that varies with the *degrees of freedom* with a minimum value of 0.[21]

χ^2 statistic—A value obtained from the χ^2 *distribution* at a given percentage point and a given *degree of freedom*.[21]

χ^2 test—A *statistic* used in testing a *hypothesis* concerning the discrepancy between *observed* and expected results.[21]

x—The observed value of a quality characteristic; specific observed values are designated $x1, x2, x3$, and so on. x is also used as a predictor value. See *input variable* or *predictor variable*.

$\bar{x}$ (pronounced X-bar)—The *average*, or *arithmetic mean*. The average of a set of n observed values is the sum of the observed values divided by n.[3]

$$\bar{x} = \frac{x_1 + x_2 + \ldots + x_n}{n}$$

$\bar{\bar{x}}$ (pronounced X-double bar)—The average for the set under consideration of the subgroup values of $\bar{x}$.[8]

$\bar{x}$ chart (average control chart)—A *variables control chart* for evaluating the *process* level in terms of *subgroup averages*.

Central line: $\bar{x}$ (if standard given, $\bar{x}_0$)

Control limits: $\bar{\bar{x}} \pm A_3\bar{s}$ or $\bar{\bar{x}} \pm A_2\bar{R}$ (if standard given, $\bar{x}_0 \pm A\,\sigma_0$ or $\bar{x}_0 \pm A_2R_0$)

Where $\bar{\bar{x}}$ is the average of the subgroup values; $\bar{s}$ is the sample *standard deviation*; A, $A2$, and $A3$ are *control chart factors*; σ_0 is the standard value of the population standard deviation; and R_0 is the standard value of the *range*. (Use the formula with $\bar{R}$ when the sample size is small; the formula with $\bar{s}$ when the sample is larger—generally >10 to 12.).[5,7,8,12]

y—y is sometimes used as an alternate to x as an observation. In such cases, $y0$, and should be substituted where appropriate. See *output variable* or *response variable*.[9]

z-test (one-sample)—Given an unknown *mean* μ, and known *variance* σ^2:

$$z = \frac{\bar{x} - u_0}{\sigma/\sqrt{n}}$$

where $\bar{x}$ is the mean of the data, μ_0 is the standard value of the *population mean*, σ is the *population standard deviation*, and n is the *sample* size.

Note: If the variance is unknown, then the *t-test* applies.[17]

Glossary References

1. ANSI/ASQ Z1.9-2003, *Sampling Procedures and Tables for Inspection by Variables for Percent Nonconforming* (Milwaukee: American Society for Quality).

2. ANSI/ASQC A1-1978, *Definitions, Symbols, Formulas, and Tables for Control Charts* (Milwaukee: American Society for Quality Control).

3. ANSI/ASQC A2-1978, *Definitions, Symbols, Formulas, and Tables for Control Charts* (Milwaukee: American Society for Quality Control).

4. ANSI/ASQC B2-1996, *Control Chart Method of Analyzing Data* (Milwaukee: American Society for Quality Control).

5. ANSI/ASQC B3-1996, *Control Chart Method of Controlling Quality During Production* (Milwaukee: American Society for Quality Control).

6. ANSI/ISO/ASQ Q9000-2000, *Quality management systems—Fundamentals and vocabulary* (Milwaukee: American Society for Quality).

7. ASQ Chemical and Process Industries Division, Chemical Interest Committee, *Quality Assurance for the Chemical and Process Industries*, 2nd ed. (Milwaukee: ASQ Quality Press, 1999).

8. ASQ Statistics Division, *Glossary and Tables for Statistical Quality Control*, 3rd ed. (Milwaukee: ASQ Quality Press, 1996).

9. ASQC Statistics Technical Committee, *Glossary and Tables for Statistical Quality Control* (Milwaukee: American Society for Quality Control, 1973).

10. Allan G. Bluman, *Elementary Statistics: A Brief Version*, 2nd ed. (New York: McGraw Hill, 2003).

11. BSR/ISO/ASQ 3534-1.3-200X, *Statistics—Vocabulary and Symbols—Part 1: Probability and General Terms* (Milwaukee: American Society for Quality Control).

12. BSR/ISO/ASQ 3534-2-200X, *Statistics—Vocabulary and Symbols—Part 2: Applied Statistics* (Milwaukee: American Society for Quality Control).

13. BSR/ISO/ASQ 3534-3-1999, *Statistics—Vocabulary and Symbols—Part 3: Design of Experiments* (Geneva, Switzerland: International Organization for Standardization, 1999).

14. David C. Hoaglin, Frederick Mosteller, and John W. Tukey, *Understanding Robust and Exploratory Data Analysis* (New York: John Wiley & Sons, 1983).

15. Samuel Kotz and Norman L. Johnson, "Process Capability Indices—A Review, 1992–2000" *Journal of Quality Technology* 34, no. 1 (2000): 2ff.

16. Minitab, Inc., *Minitab, Release 13* (State College, PA: Minitab, 2001).

17. Douglas C. Montgomery, *Introduction to Statistical Quality Control*, 4th ed. (New York: John Wiley & Sons, 2001).

18. Lloyd S. Nelson, "Notes on the Shewhart Control Chart," *Journal of Quality Technology* 31, no. 1 (1999).

19. NIST/SEMATECH e-Handbook of Statistical Methods http://www.itl.nist.gov/div898/handbook/, February 15, 2004.

20. Barbara F. Ryan, Brian L. Joiner, and Thomas A. Ryan, *Minitab Handbook*, 2nd ed. (Boston: Duxbury Press, 1985).

21. Team definition developed by *Glossary and Tables for Statistical Quality Control,* Fourth Edition authors.

22. John W. Tukey, *Exploratory Data Analysis* (Reading, MA: Addison-Wesley, 1977).

23. Paul F. Velleman and David C. Hoaglin, *Applications, Basics, and Computing of Exploratory Data Analysis* (Boston: Duxbury Press, 1981).

References

Anderson, David R., Dennis J. Sweeney, and Thomas A. Williams. 2011. *Statistics for Business and Economics*. 11th ed. Mason, OH: South-Western College Learning.

Benbow, Donald W., and T. M. Kubiak. 2009. *The Certified Six Sigma Black Belt Handbook*. Milwaukee, WI: ASQ Quality Press.

Besterfield, Dale A. 1998. *Quality Control*. 5th ed. Upper Saddle River, NJ: Prentice Hall.

Borror, Connie M., ed. 2009. *The Certified Quality Engineer Handbook*. 3rd ed. Milwaukee, WI: ASQ Quality Press.

Boslaugh, Sarah, and Paul Andrew Watters. 2008. *Statistics in a Nutshell: A Desktop Quick Reference*. Sebastopol, CA: O'Reilly Media.

Box, George E. P., J. Stuart Hunter, and William G. Hunter. 2005. *Statistics for Experimenters: Design, Innovation, and Discovery*. 2nd ed. Hoboken, NJ: John Wiley & Sons.

Breyfogle, Forrest III. 2008. *Integrated Enterprise Excellence Volume III—Improvement Project Execution: A Management and Black Belt Guide for Going Beyond Lean Six Sigma and the Balanced Scorecard*. Austin, TX: Citius Publishing.

Burrill, Claude W., and Johannes Ledolter. 1999. *Achieving Quality Through Continual Improvement*. New York: John Wiley & Sons.

Chrysler, Ford, and General Motors. 2010. *Measurements System Analysis*. 4th ed. Automotive Industry Action Group.

Clarke, G. M., and D. A. Cooke. 1992. *Basic Course in Statistics*. 3rd ed. Kent, England: Edward Arnold.

Diamond, William J. 1981. *Practical Experimental Designs: For Engineers and Scientists*. Belmont, CA: Wadsworth.

George, Michael L., David Rowlands, Mark Price, and John Maxey. 2005. *The Lean Six Sigma Pocket Tool Book*. New York: McGraw-Hill.

Goldberg, Samuel. 1986. *Probability: An Introduction*. New York: Dover Publications.

Gryna, Frank M. 2001. *Quality Planning and Analysis*. 4th ed. New York: McGraw-Hill.

Hinton, Perry R. 2004. *Statistics Explained: A Guide for Social Science Students.* 2nd ed. New York: Routledge.

Johnson, Allen G. 1988. *Statistics.* New York: Harcourt Brace Jovanovich.

Johnson, Richard A., Gouri K. Bhattacharyya. 2010. *Statistics: Principles and Methods.* Hoboken, NJ: John Wiley & Sons.

Kenney, James M. 1988. "Hypothesis Testing: Guilty or Innocent." *Quality Progress* 21, no. 1: 55–58.

Lawson, John, and John Erjavec. 2001. *Modern Statistics for Engineering and Quality Improvement.* Pacific Grove, CA: Wadsworth Group.

McClave, James T., and Frank H. Dietrich II. 1991. *Statistics.* 5th ed. New York: Dellen Publishing Company.

Meek, Gary E., Howard L. Taylor, Kenneth A. Dunning, and Keith A. Klafehn. 1987. *Business Statistics.* Boston: Allyn and Bacon.

Metcalfe, Andrew W. 1994. *Statistics in Engineering: A Practical Approach.* London: Chapman and Hall.

Montgomery, Douglas C., and George C. Runger. 1999. *Applied Statistics and Probability for Engineers.* 2nd ed. New York: John Wiley & Sons.

Montgomery, Douglas C., George C. Runger, and Norma F. Hubele. 2001. *Engineering Statistics.* 2nd ed. New York: John Wiley & Sons.

Pries, Kim H. 2009. *Six Sigma for the New Millennium.* 2nd ed. Milwaukee, WI: ASQ Quality Press.

Shankar, Rama. 2009. *Process Improvement Using Six Sigma: A DMAIC Guide.* Milwaukee, WI: ASQ Quality Press.

Simmons, J. P., L. D. Nelson, and U. Simonsohn. 2011. "False-Positive Psychology: Undisclosed Flexibility in Data Collection and Analysis Allows Presenting Anything as Significant." *Psychological Science* 22, no. 11: 1359–66.

Warner, Rebecca M. 2008. *Applied Statistics: From Bivariate Through Multivariate Techniques.* Thousand Oaks, CA: Sage Publications.

Weimar, Richard C. 1993. *Statistics.* 2nd ed. Dubuque, IA: Wm. C. Brown Publishers.

Witte, Robert S. 1993. *Statistics.* 4th ed. New York: Harcourt Brace College Publishers.

Zilak, Stephen T. 2008. "Guinnessometrics: The Economic Foundation of 'Student's' t." *Journal of Economic Perspectives* 22, no. 4: 199–216.

Index

Note: Page numbers followed by *f* or *t* refer to figures or tables, respectively.

The Knowledge Center
www.asq.org/knowledge-center

Learn about quality. Apply it. Share it.

ASQ's online Knowledge Center is the place to:

- Stay on top of the latest in quality with Editor's Picks and Hot Topics.
- Search ASQ's collection of articles, books, tools, training, and more.
- Connect with ASQ staff for personalized help hunting down the knowledge you need, the networking cpportunities that will keep your career and organization moving forward, and the publishing opportunities that are the best fit for you.

Use the Knowledge Center Search to quickly sort through hundreds of books, articles, and other software-related publications.

www.asq.org/knowledge-center

PUBLICATIONS

ASQ

The Global Voice of Quality™

Ask a Librarian

Did you know?

- The ASQ Quality Information Center contains a wealth of knowledge and information available to ASQ members and non-members

- A librarian is available to answer research requests using ASQ's ever-expanding library of relevant, credible quality resources, including journals, conference proceedings, case studies and Quality Press publications

- ASQ members receive free internal information searches and reduced rates for article purchases

- You can also contact the Quality Information Center to request permission to reuse or reprint ASQ copyrighted material, including journal articles and book excerpts

- For more information or to submit a question, visit **http://asq.org/knowledge-center/ ask-a-librarian-index**

Visit www.asq.org/qic for more information.

TRAINING   CERTIFICATION   CONFERENCES   MEMBERSHIP   **PUBLICATIONS**

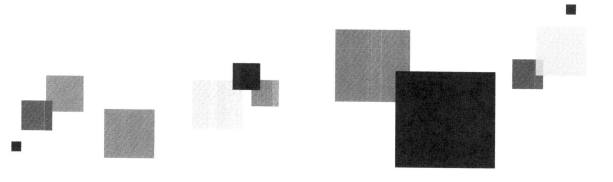

Belong to the Quality Community!

Established in 1946, ASQ is a global community of quality experts in all fields and industries. ASQ is dedicated to the promotion and advancement of quality tools, principles, and practices in the workplace and in the community.

The Society also serves as an advocate for quality. Its members have informed and advised the U.S. Congress, government agencies, state legislatures, and other groups and individuals worldwide on quality-related topics.

Vision

By making quality a global priority, an organizational imperative, and a personal ethic, ASQ becomes the community of choice for everyone who seeks quality technology, concepts, or tools to improve themselves and their world.

ASQ is...

- More than 90,000 individuals and 700 companies in more than 100 countries

- The world's largest organization dedicated to promoting quality

- A community of professionals striving to bring quality to their work and their lives

- The administrator of the Malcolm Baldrige National Quality Award

- A supporter of quality in all sectors including manufacturing, service, healthcare, government, and education

- YOU

Visit www.asq.org for more information.

TRAINING   CERTIFICATION   CONFERENCES   MEMBERSHIP   **PUBLICATIONS**   The Global Voice of Quality™

ASQ Membership

Research shows that people who join associations experience increased job satisfaction, earn more, and are generally happier*. ASQ membership can help you achieve this while providing the tools you need to be successful in your industry and to distinguish yourself from your competition. So why wouldn't you want to be a part of ASQ?

Networking

Have the opportunity to meet, communicate, and collaborate with your peers within the quality community through conferences and local ASQ section meetings, ASQ forums or divisions, ASQ Communities of Quality discussion boards, and more.

Professional Development

Access a wide variety of professional development tools such as books, training, and certifications at a discounted price. Also, ASQ certifications and the ASQ Career Center help enhance your quality knowledge and take your career to the next level.

Solutions

Find answers to all your quality problems, big and small, with ASQ's Knowledge Center, mentoring program, various e-newsletters, *Quality Progress* magazine, and industry-specific products.

Access to Information

Learn classic and current quality principles and theories in ASQ's Quality Information Center (QIC), *ASQ Weekly* e-newsletter, and product offerings.

Advocacy Programs

ASQ helps create a better community, government, and world through initiatives that include social responsibility, Washington advocacy, and Community Good Works.

Visit www.asq.org/membership for more information on ASQ membership.

*2008, The William E. Smith Institute for Association Research

TRAINING   CERTIFICATION   CONFERENCES   **MEMBERSHIP   PUBLICATIONS**